Studies in Big Data

Volume 63

The series "Studies in Big Data" (SBD) publishes new developments and advances in the various areas of Big Data- quickly and with a high quality. The intent is to cover the theory, research, development, and applications of Big Data, as embedded in the fields of engineering, computer science, physics, economics and life sciences. The books of the series refer to the analysis and understanding of large, complex, and/or distributed data sets generated from recent digital sources coming from sensors or other physical instruments as well as simulations, crowd sourcing, social networks or other internet transactions, such as emails or video click streams and other. The series contains monographs, lecture notes and edited volumes in Big Data spanning the areas of computational intelligence including neural networks, evolutionary computation, soft computing, fuzzy systems, as well as artificial intelligence, data mining, modern statistics and Operations research, as well as self-organizing systems. Of particular value to both the contributors and the readership are the short publication timeframe and the world-wide distribution, which enable both wide and rapid dissemination of research output.

** Indexing: The books of this series are submitted to ISI Web of Science, DBLP, Ulrichs, MathSciNet, Current Mathematical Publications, Mathematical Reviews, Zentralblatt Math: MetaPress and Springerlink.

More information about this series at http://www.springer.com/series/11970

Prasant Kumar Pattnaik ·
Raghvendra Kumar · Souvik Pal ·
S. N. Panda
Editors

IoT and Analytics for Agriculture

Springer

Editors
Prasant Kumar Pattnaik
School of Computer Engineering
Kalinga Institute of Industrial Technology
Bhubaneswar, Odisha, India

Raghvendra Kumar
Department of Computer Science
and Engineering
Laxmi Narayan College of Technology
Jabalpur, Madhya Pradesh, India

Souvik Pal
Department of Computer Science
and Engineering
JIS College of Engineering
Nadia, West Bengal, India

S. N. Panda
Chitkara University
Chandigarh, Punjab, India

ISSN 2197-6503 ISSN 2197-6511 (electronic)
Studies in Big Data
ISBN 978-981-13-9176-7 ISBN 978-981-13-9177-4 (eBook)
https://doi.org/10.1007/978-981-13-9177-4

This Springer imprint is published by the registered company Springer Nature Singapore Pte Ltd.
The registered company address is: 152 Beach Road, #21-01/04 Gateway East, Singapore 189721, Singapore

Preface

This edited book aims to bring together leading academic scientists, researchers, and research scholars to exchange and share their experiences and research results on all aspects of wireless IoT and analytics for agriculture. It also provides a premier interdisciplinary platform for researchers, practitioners, and educators to present and discuss the most recent innovations, trends, and concerns as well as practical challenges encountered and solutions adopted in the fields of IoT and analytics for agriculture. The book is organized into 12 chapters.

Chapter "Integrating Big Data Practices in Agriculture", focuses on the integration of Big Data practices in agronomical practices, supply chain operation, and consumers' feedback, by using different Big Data approaches. This chapter would help in understanding the multifaceted concept of Big Data in various agricultural practices.

Chapter "Solar-Powered Automated IoT-Based Drip Irrigation System", presents wireless sensors that were used to analyze the moisture level of the field and to automate the irrigation process. An automated irrigation system refers to the operation of the system with a minimum of manual intervention besides the surveillance. All types of irrigation process like drip, sprinkler, or surface can be automated with the help of timers, sensors, or computers or mechanical appliances. In this automated system, some standard data regarding the moisture level of a field is predefined for a specific crop. The sensor first transmits the data through a microcontroller to the server after a specific time interval. If the moisture level of the field is below the standard value of the database, the microcontroller will trigger the pump for watering the plants till the predefined moisture level. If the moisture level of the field reaches the threshold limit, then the pump will automatically stop watering through microcontroller.

Chapter "IoT Foundations and Its Application", deals with the concept of Internet of things (IoT), the applications of IoT, the technologies of IoT, and the need of IoT.

Chapter "e-Device for the Protection of Agricultural Land from Elephant Attacks in Odisha: A Review", presents an approach to reduce crop damage by the elephants on agricultural lands. This proposed work is comprised of Internet of things and cloud-based technology with high-resolution camera to capture the images and process it and compare it with already stored images on cloud with the help of image processing algorithm, and after detection, it intimates the corresponding authority. This approach system would be fully automatic, and it is a low-cost device and affordable for large area coverage and ultimately this will decrease the loss of lives.

Chapter "The Impact of Irrigation on Agricultural Productivity in the Bolpur Sub Division, West Bengal", explores the fact that ever-increasing population leads to more intensive agriculture which needs more irrigation facilities, along with other inputs. Agricultural productivity of any region is closely associated with a number of physical (like relief, climate, soil, water, etc.), socioeconomic, political, institutional, and organizational factors of that region. Higher-level application of chemical fertilizers and pesticide, timely and adequate supply of irrigation, and higher-level supervision help to have an upper-level productivity of crops in the study area.

Chapter "Big Data, Climate Smart Agriculture, and India–Africa Relations: A Social Science Perspective", analyzes the possibilities of collaboration across agriculture and allied activities that can benefit both Indian and African economies by integrating Big Data into climate smart agriculture, which thereby increases agricultural productivity as well as efficiency in the utilization of resources. It takes an exploratory and descriptive research, with the intention of filling the void of lack of literature in social sciences.

Chapter "IoT-Enabled Agricultural System Applications, Challenges and Security Issues", aims at agricultural applications where it utilizes modern technologies that benefit the farmers with decision tools and reduce manual laboring cost. The seamless integration of products, knowledge, and services through IoT maximizes the volume of productivity, product quality, and profit of business. Even though current surveys on the IoT in agriculture focuses on the challenges, constraints, benefits, and pitfalls for large scale in the agricultural food sector, all are presented in isolation to each other. So, keeping all in these in mind, a brief discussion on challenges, benefits, constraints, future trends, and security issues is presented in this chapter.

Chapter "Land Suitability Analysis for Peri-urban Agriculture Using Multi-criteria Decision Analysis Model and Crop Condition Monitoring Methods: A Case Study of Kolkata Metropolitan Area", discusses multi-criteria decision analysis model and applies crop condition monitoring method to find out the suitable zones of peri-urban agriculture around Kolkata city. This model was tested on the Kolkata Metropolitan Area (KMA) using various criteria like land use, digital elevation model (DEM), water facility, road and market facility, etc. For analyzing the land use of KMA, supervised classification methods has been used, and to find out the agricultural area, crop condition monitoring method (e.g., NDVI, i.e., Normalized Difference Vegetation Index) has been used by using remote sensing images. For the analysis of water, road, and market facilities in various areas of KMA, census data has been used. Multi-criteria decision analysis model revealed that northwest, central-east, southeast, and northern zones of KMA are the most suitable for peri-urban agriculture. Finally, it can be said that this model is able to allocate the suitable land for the peri-urban agriculture very precisely. This model will help the urban, peri-urban planners, policy-makers, and decision-makers for taking action on various decisions at different levels.

Chapter "Security and Privacy Issues in Wireless Sensor Networks", discusses sensor node which is used to collect the information from the environment and transmit the data to the control unit. Due to a large number of collection and transmission, should need to protect from hacker. WSNs need to provide security in all aspects of their architecture reference model. In this chapter main contribution includes architecture, attacks, and countermeasures.

Chapter "A Design of IoT-Based Agricultural System for Optimal Management", designed IoT-based agricultural system using data mining and communication protocol/network for optimal management of agriculture. IoT-based technology makes nonstop message between object and things or between things/object and humans. Data mining and communication protocol will be used as prediction/inference of the knowledge from the hidden data and make the optimal travel of the data with security and recovery, respectively. Therefore, in this chapter, an effort has been made to design an IoT-based agricultural system and discuss the problem and implication of the implementation of these techniques.

Chapter "Smart and Sustainable Agriculture Through IoT Interventions: Improvisation, Innovation, and Implementation—An Exploratory Study", exemplifies by the glimpse of application like farming based on weather projection, real-life count of agriculture produces, real-life estimation for loss due to ability or expiry, irrigation issues, controlling of infrastructure support for farming activities from a distant location, census of cattle, etc. In fact, the concept of IoT is still in nascent stage in India. There are vast opportunities of IoT application in the country since India is primarily an agrarian society and around 60% population are engaged in this profession which contributes around 17% of share in GDP and feeding the elephantine population of the country. This paper would study various sparks of IoT system, its versatile application worldwide, and possible intervention in India, particularly in agricultural activities. The paper would explore innovative modeling for IoT integration in agriculture system and its ease of implementation globally with emphasis on Indian subcontinent.

We are sincerely thankful to Almighty for supporting and standing at all times with us, whether it is good or tough times and given ways to concede us. Starting from the call for chapters till the finalization of chapters, all the editors have given their contributions amicably, which is a positive sign of significant teamwork. The editors are sincerely thankful to all the members of Springer, especially

Mr. Aninda Bose and Ms. Shilpa, for providing the constructive inputs and allowing an opportunity to edit this important book. We are equally thankful to reviewers who hail from different places in and around the globe and shared their support and stand firm toward quality chapter submission.

Bhubaneswar, India	Prasant Kumar Pattnaik
Jabalpur, India	Raghvendra Kumar
Nadia, India	Souvik Pal
Chandigarh, India	S. N. Panda

About This Book

A major challenge in agriculture is to cultivate produce in the farm and deliver it to the end consumers with the best possible price and best possible quality. Currently, all over the world, it is found that around 50% for the farm produce never reach the end consumer due to wastage and suboptimal prices. The audience will get the solution to reduce the transport cost, predictability of prices on the past data analytics, and the current market conditions, reduce number of middle hops and agents between the farmer and the end consumer using IoT-based solution. Again, the demand by consumption of agricultural products could be predicted quantitatively; however, the variation of harvest and production by the change of farm's cultivated area, weather change, disease, insect damage, etc. could be difficult to be predicted, so that the supply and demand of agricultural products have not been controlled properly. To overcome it, this edited book designed the IoT-based monitoring system to analyze crop environment and the method to improve the efficiency of decision making by analyzing harvest statistics. The book also intended to attract the audience who work in climate changes. Climate change and rainfall have been erratic over the past decade. Due to this in the recent era, climate-smart methods called as smart agriculture is adopted by many Indian farmers. Smart agriculture is an automated and directed information technology with the IoT.

Key Features

1. Addresses the complete functional framework workflow in IoT-enabled agrosystem.
2. Explores basic and high-level concepts, thus serving as a manual for those in the industry while also helping beginners to understand both basic and advanced aspects in IoT-based agriculture-related issues.
3. Based on the latest technologies and covering the major challenges, issues, and advances in IoT agrosystem.
4. Exploring intelligent field monitoring and automated system through IoT ecosystem and its implications to the real world.
5. Explains the concepts of smart monitoring and irrigation in IoT-based agricultural system for the betterment of the smarter humanity.
6. Intelligent data processing and sensor technologies in IoT-enabled agriculture system.
7. Exploring the predictive analysis system, crop monitoring, and weather data-enabled analysis in IoT agrosystem.
8. Exploring data acquisition and case studies related to data-intensive technologies in farming-based Internet of things.

Contents

About the Editors

Prasant Kumar Pattnaik Ph.D (Computer Science), Fellow IETE, Senior Member IEEE is a Professor at the School of Computer Engineering, KIIT Deemed University, Bhubaneswar. He has more than a decade of teaching and research experience. Dr. Pattnaik has published numbers of Research Papers in peer-reviewed International Journals and Conferences. He also published many edited book volumes in Springer and IGI Global Publication. His areas of interest include Mobile Computing, Cloud Computing, Cyber Security, Intelligent Systems and Brain Computer Interface. He is one of the Associate Editor of Journal of Intelligent & Fuzzy Systems, IOS Press and Intelligent Systems Book Series Editor of CRC Press, Taylor Francis Group.

Raghvendra Kumar is working as Associate Professor in Computer Science and Engineering Department at L.N.C.T Group of College Jabalpur, M.P. India, and Serving as Director of IT and Data Science Department, Vietnam Center of Research in Economics, Management, Environment (VCREME) - Branch VCREME One Member Company Limited, Vietnam. He received B. Tech. in Computer Science and Engineering from SRM University Chennai (Tamil Nadu), India, M. Tech. in Computer Science and Engineering from KIIT University, Bhubaneswar, (Odisha) India and Ph.D. in Computer Science and Engineering from Jodhpur National University, Jodhpur (Rajasthan), India. He serves as Series Editor Internet of Everything (IOE): Security and Privacy Paradigm publishes by CRC press, Taylor & Francis Group, USA and Bio-Medical Engineering: Techniques and Applications, Publishes by Apple Academic Press, CRC Press, Taylor & Francis Group, USA. He has published number of research papers in international journal (SCI/SCIE/ESCI/Scopus) and conferences including IEEE and Springer as well as serve as organizing chair (RICE-2019), volume Editor (RICE-2018), Keynote speaker, session chair, Co-chair, publicity chair, publication chair (NGCT-2017), advisory board, Technical program Committee members in many international and national conferences and serve as guest editors in many special issues from reputed journals (Indexed By: Scopus, ESCI). He also published 11 chapters in edited book published by IGI Global, Springer and Elsevier. He also received best paper award in IEEE Conference 2013 and Young Achiever Award-2016 by IEAE Association for his research work in the field of distributed database. His researches areas are Computer Networks, Data Mining, cloud computing and Secure Multiparty Computations, Theory of Computer Science and Design of Algorithms. He authored and Edited 17 computer science books in field of Internet of Things, Data Mining, Biomedical Engineering, Big Data, Robotics, Graph Theory, and Turing Machine by IGI Global Publication, USA, IOS Press Netherland, Springer, Elsevier, CRC Press, USA, S. Chand Publication and Laxmi Publication. He is Managing Editor in International Journal of Machine Learning and Networked Collaborative Engineering (IJMLNCE) ISSN 2581-3242.

Souvik Pal Ph.D., MCSI, MCSTA/ACM, USA; MIAENG, Hong Kong; MIRED, USA; MACEEE, New Delhi; MIACSIT, Singapore; MAASCIT, USA is working as Assistant Professor at the Department of Computer Science and Engineering in JIS College of Engineering, Kalyani, India. Dr. Pal has received his both Masters Degree and Doctorate Degree from KIIT University, Bhubaneswar, India. Dr. Pal has published several research papers in peer-reviewed International journals and conferences (SCOPUS, ESCI). He has authored a book on computer science in the field of Cloud Computing. He was appointed in many conferences as Session chair, reviewer, and track co-chair. Dr. Pal also serves as Editorial and International Advisory board member for many journals and conferences. His research area includes Cloud Computing, Big Data, Internet of Things, and Data Analytics.

S. N. Panda born in Jharsuguda, Odisha on 19th August 1969. He completed his B.Sc. (Hons) with First class Distinction from Laxmi Narayan College, Jharsuguda, then he pursued his M.Sc. from G.J.University, Hissar and Ph.D. (Computer Science) from Kurukshetra University, Haryana. He joined as Lecturer in S.A.Jain (PG) College, Ambala. In the year 2007 he joined as Professor and Principal Regional Institute of Management and Technology, Punjab and is presently working as Professor and Director Research, Chitkara University, Punjab. He has more 25 years of Teaching/ Administration and Research experience. Guided 7 Ph. D. Computer Science and 13 M.Phil Scholar. He is now working towards development of innovative technologies and product based on Internet of things and Cloud Computing. He is expertized in Cyber security, Networking, Advanced Computer Network, Machine Learning and Artificial Intelligence. He has filed 8 patents, 55 Intenational publications in the relevant area and involved in Internet of things healthcare devices like Portable Intensive Care Unit, Digital Laryngoscope etc. He has developed the prototype of Smart Portable Intensive Care Unit through which doctor can provide the immediate virtual medical assistance to emergency cases in ambulance and won prestigious Millennium Alliance Award from FICCI in 2017 and seed funding for his project. He is also working on a project "Cyber Technology Communication for Women Safety" which is funded by Ministry of Science and Technology, Govt. Of India and another project "Remote Vital Information and Surveillance System for Elderlyand Disabled Persons" which is again funded by Ministry of Science and Technology, Govt. Of India. He has been bestowed with "Adroit Researcher Award" for the outstanding contribution in the field of Education and Research, during the "International Conference on Interdisciplinary Research for Sustainable Development (IRSD-2016)" in NITTTR Chandigarh and COSMIC Outstanding Researcher Award from COSMIC Journal, Thailand. He is also proud recipient of Teacher's Excellence Award from Chitkara University, Punjab INDIA. He received Certificate of Appreciation as Productive Member of International Reviewer Board, Informing Science, Vilnius, Lithuania (Europe) 2014 and Appreciation Award by ISI Florida, USA for excellent work as Member of International Board of Reviewer 2015. He represented Chitkara University in TiECON 2017, in Silicon Valley, California, USA.

Integrating Big Data Practices in Agriculture

Jolly Masih and Rajkumar Rajasekaran

Abstract The world is facing shortage of food supply due to lack of integration and utilization of technology in agriculture. Huge information available online about cultivation using drones, details about production and consumption of fertilizer, crop productivity and production data could be used efficiently to make farming practices better and more efficient. Big Data provides a high volume, speed and assortment required for particular innovation and explanatory strategies for efficient agriculture operation right from farm cultivation to marketing. In this chapter, we have laid focus on integration of Big Data practices in agronomical practices, supply chain operation and consumers' feedback, by using different Big Data approaches. This chapter would help in understanding the multifaceted concept of Big Data in various agricultural practices.

Keywords Big data · Agriculture · Sentiment analysis · Internet of things (IoT) · Fertilizer · Gluten-free

1 Introduction

"The world will not always need engineers but will always need farmers to grow food and to feed the humans". The fusion of technology with farming practices could be a good solution to meet the growing world demand of food and nutrition. The world is facing shortage of food supply due to lack of integration and utilization of technology in agriculture, which is also accompanied by other problems like environmental change, land and water imperatives, increment in urbanization, ecological corruption , changing wage and diets. Integration of Big Data practices in agriculture

J. Masih (✉)
Erasmus School of Economics, Erasmus University, Rotterdam, The Netherlands
e-mail: jolly.iabm@gmail.com

R. Rajasekaran
School of Computer Science and Engineering, Vellore Institute of Technology, Vellore, Tamil Nadu, India
e-mail: vitrajkumar@gmail.com

P. K. Pattnaik et al. (eds.), *IoT and Analytics for Agriculture*, Studies in Big Data 63,
https://doi.org/10.1007/978-981-13-9177-4_1

can be a possible solution to the challenges and problems mentioned above. Huge information available online about cultivation using drones, details about production and consumption of fertilizer, crop productivity and production data could be used efficiently to make farming practices better and more efficient. Big Data has such a boundless effect in the farming business that it is difficult to pinpoint every one of its belongings, harder still to anticipate what transformation it might bring. Big Data provides a high volume, speed and assortment to require particular innovation and explanatory strategies for efficient agriculture operation right from farm cultivation to marketing. Enormous information is being utilized to give knowledge about cultivating tasks, drive constant operational choices, and upgrade business forms for diversion and expansion.

Horticultural Big Data is accepted to be very unstructured and heterogeneous in nature [1]. The matter of heterogeneity of information originates from the way that it would be gathered by different ways. Information gathered from the agriculturists, field or the homestead incorporate data on determinations of harvests, seeds, plant assortment, planting pattern, compost, showering, materials, yields, in-season symbolism, soil types, climate, and different practices. There are as a rule three classifications of information age [2, 3]: (i) process intervened (PM), (ii) mama chine produced (MG) and (iii) human sourced (HS).

Process Mediated data is also called traditional business data. It is a data of agricultural processes involving activities like purchase of agriculture inputs, seeds, fertilizer application, maintaining record of the orders and events, etc. It is usually highly structured and organized in nature and easy to analyze. Traditional business data are the huge amount IT managed and processed data, available in both operational and business information systems, usually structured and stored in relational database systems [4].

Machine-generated data is obtained from the satellite or remote sensing technique or drones. The data is collected by help of sensors and other smart machines which help to measure and record farming different processes. Machine-generated data is now a days used as "Internet of Things" (IoT). Its well-structured nature is suitable for computer processing, but its size and speed is beyond traditional approaches. IoT has been used for Smart Farming. Drones with well-equipped modern machinery including infrared cameras, GPS technology, weather sensors, pets or disease detectors are used in smart farming these days [5]. In animal husbandry, smart dairy farms are used in activities like feeding managing the cows, getting the milk from them, maintaining the daily health record [6]. IoT is also used in precision farming to manage the record of each seedling, plot, soil and water conditions, etc. [7]. With these new technologies, others data formats like sounds or images or wavelengths or graphical patterns can also be used in the analysis [8].

Human-sourced data is obtained from data published in books, articles, journals and works of art in form of photographs, graphics, text, audio and video. Human sourced information is now mostly digitalized and stored everywhere from personal computers to social networks. Human-sourced data are generally not well-structured, loosely structured, messy, unorganized and often ungoverned. In the context of Big Data and Smart Farming, human sourced or data from social media could be used efficiently to study the consumer behavior, consumer sentiments and product feedback. Various social media platforms like Facebook, Twitter and Instagram, etc. could be used to get the data [4].

Table 1 provides an overview of current Big Data applications in relation to agriculture sector. Examples of Big Data applications/aspects in agriculture.

Feasible and proper utilization of Big Data is required for progressive advancement in agriculture and related fields like agribusiness, food technology, nutrition, horticulture, etc.

Incorporation of Big Data rehearses in farming could be fruitful if funding firms identified with horticulture turns into a piece of this evolving pattern. This will thought process ranchers to actualize Big Data rehearses in farming and partnered fields. Barely any instances of funding firms putting resources into farming innovation organizations, for example, Blue River Technology, a business concentrating on the utilization of PC vision and mechanical autonomy in agribusiness [9]. The new players of Big Data application in Smart Farming are tech organizations that were generally not all around associated with horticulture. For instance, Japanese innovation firms, for example, Fujitsu are helping agriculturists with their cloud-based cultivating frameworks [5]. Fujitsu gathers information identified with atmosphere conditions, precipitation, mugginess, soil temperatures from a system of cameras

Table 1 Big data applications

Big Data technique used	Arable	Livestock	Horticulture	Fishery
Sensing and monitoring	Robotics and sensors	Biometric sensing, GPS tracking	Robotics and sensors (temperature, humidity, CO_2, etc.), greenhouse computers	Automated identification systems (AIS)
Analysis and planning	Seeding, planting, soil typing, crop health, yield modeling	Breeding, monitoring	Lighting, energy management	Surveillance, monitoring
Smart control	Precision farming	Milk robots	Climate control, precision control	Surveillance, monitoring
Big Data in the cloud	Weather/climate data, yield data, soil types, market information, agricultural census data	Livestock movements	Weather/climate, market information, social media	Market data Satellite data

Source [4]

and sensors introduced the nation over to help agriculturists in Japan better deal with its products and costs. Organization additionally advances information professional processing masters to end up accomplices of makers as Big Data conveys on its guarantee to on a very basic level change the aggressiveness of makers. This methodology has given proficiency in cultivating rehearses.

To accept the benefit of open doors for Big Data in the agribusiness segment, huge horticulture organizations, for example, Monsanto, Mahindra and Mahindra and John Deere have spent immense measure of cash on innovations that utilization point by point information on soil type, seed assortment, N:P:K piece of the field and climate to enable agriculturists to cut expenses and increment yields. Different players incorporate different quickening agents, hatcheries, investment firms and corporate endeavor reserves (Monsanto, DuPont, Syngenta, Bayer, DOW and so forth) [10].

Monsanto being a market pioneer has been concentrating on different huge information approaches, investigation and precision over the entirety of its business lines, from atmosphere forecast to hereditary designing and its execution. Monsanto has been exceptionally steady in its endeavors to influence an ever increasing number of ranchers to receive its cloud benefits by telling the advantages of Big Data in agribusiness. Monsanto trusts that agriculturist's advantage most when they permit the company to break down their information—alongside that of different ranchers—to enable them to locate the best answers for each fix of land [11].

Expanding association and commitment of corporate with Big Data application and agribusiness over that world gives a chance to, new businesses to give arrangements over the esteem chain, from foundation and sensors the distance down to programming that deals with the numerous floods of information from over the ranch. As the agri-tech space warms up, an expanding number of little tech new companies are propelling items giving their greater partners a keep running for their cash.

In this section, we would focus on utilizing distinctive Big Data information for smart agriculture. Chapter would put light on various parts of Big Data which could be linked with agribusiness for welfare of farmers and consumers. We have thought about the following parameters:

- Using Big Data to improve Agronomical practices
- Using online data for betterment of farmers
- Use of Social media analytics in Agriculture through following techniques.
  - Sentimental analytics
  - Perceptual mapping or modeling
  - Online portals
  - Feedback model (Discriminate Analysis, Factor Analysis) (Behavior of the customer).

Enormous Data applications may be linked to two extremes of agribusiness field: firstly, incorporation of Big Data practices in the production and cultivation process, e.g., improved cultivation practices using drowns, increasing soil and water efficiency, predicting fertilizer consumption of any crop, etc.; another in which Big Data practices could be linked with marketing side of agribusiness, e.g., to predict the sales on coming months or years, create advertisements using customer sentiments, efficient product positioning and categorization, etc.

Agriculture has been the lifestyle for thousands of years but it is one of the least focused segments for most of the countries of the world. Many places of the world like South-East Asia, Africa, South America and a few sections of Southern Europe still use old and conventional farming methods which includes the use of bullock plough, axes, sickles and other old fashioned tools. Due to lack of integration of technology in farming practices, farmers are still poor in many countries and even commit suicides.

The world needs the era of Smart Farming where farming practices would be merged with Big Data practices to bring more accuracy and advancement in agribusiness operations. The following are examples of Internet of Agricultural Things (IOAgT):

- Soil Health, Water, Nutrients
- Pests/Control
- Energy and Waste Management
- Traceability and Tracking
- Processing
- Inspection, Ripeness
- Transportation
- Storage and Inventory
- Retailer management
- Smart Refrigerators.

Talking about few examples of integration of Big Data practices in agriculture and its successful analytical implication; the total world production of groundnut in 2012–13 was approximately 37.19 million tons with India contributing about 9.47 million metric tons. India is the second largest producer of groundnuts in the world and it was forecasted using AREMA and Big Data practices that about 65.57 lakh tons of groundnut could be produced by the end of 2014–2015 production year [12]. This conveys to the current necessities of data analysis and Big Data in the field of agriculture. Studies like yield sustainability in India using SARIMA helped in provoking thought about how important are the Big Data Practices in agriculture and related fields [13].

This kind of technology needs proper analysis for their outputs. The Supply chain efficiency could be maximized using Big Data principles. A study conducted in Ghana about digital integration of agricultural markets of the country, indicated that Ajumako-Kasoa markets, Swedru-Cape Coast markets and Ajumako-Swedru markets experienced market growth after their digital integration and the market efficiency was also increased [14, 15].

The IOT services generate a lot of data for processing which when cleaned and used could give vital information. "We are drowning in information, while starving for wisdom. The world henceforth will be run by synthesizers, people able to put together the right information at the right time, think critically about it, and make important choices wisely" are the words said by a Pulitzer Prize winner on the importance of Big Data. There has been recent influx of research enthusiasts gravitating to this field [16]. A recent study conducted to predict mustard prices in Rajasthan (India), found that Big Data models with upto 95% of accuracy could be developed for successful sales prices prediction [17]. This opens up the question for breakthrough opportunities which could be generated using Big Data in agriculture. It has also been studied that VAR modeling introduction into the Big Data methods pre-existing could unveil the hidden potential in lot of ways [18]. Similar studies were conducted to find the utilization of Big Data practices in improving the rice quality based on consumed based sentiment analysis [19]. Other study included successful pigeon pea marketing in Southwest Nigeria using price variability, co-integration and exogeneity of locally produced agri-based products [20].

This information opens up several avenues of Big Data Handling:

- Analytics
- Informatics
- Evidence-Based Tools
- Meta-Analysis and Synthesis
- Complex Systems
- Computational Sciences
- Data Engineering, Mining
- Cloud Computing—Implementation and Evaluation
- Data Security and Privacy
- Predictive Modeling
- Data Visualization
- Decision Analytics, Machine Learning
- Network Science
- Spatial Analytics.

A few key issues related to integration of Big Data practices in agriculture are [4]:

- Data quality—The quality and amount of information changes as indicated by the sources and stages from where information has been taken. It is a key issue in homestead the executives data frameworks, yet is all the more difficult with huge, constant information;
- Intelligent preparing and examination—For Big Data this is likewise all the more difficult due to the vast measure of frequently unstructured, at times unsourced and mysterious, heterogeneous information which requires a savvy transaction between gifted information researchers and space specialists;
- Sustainable incorporation of Big Data sources—Integration of a wide range of information sources is challenging. Information could originate from field identified with yield condition, utilization of compost and so forth or could originate from social media about inclination of any sustenance items or item criticism and so on.; but since this is significant for your plan of action this must be done in a maintainable way;

- Developing a reasonable plan of action with astutely coordinated Big Data approaches for cultivating network Business models that are sufficiently appealing for arrangement suppliers however that likewise empower a decent amount between the distinctive partners; Openness of stages that will quicken arrangement development and development when all is said in done yet additionally engage agriculturists in their situation in supply chains.

In this chapter, we have laid focus on integration of Big Data practices in agronomical practices, supply chain operation and consumers' feedback, by using different Big Data approaches. This chapter would help in understanding the multifaceted concept of Big Data in various agricultural practices. The details of each approach are mentioned below.

2 Using Big Data to Improve Agronomical Practices

Big Data has long been tried and tested using different kinds of models and methods for the improvement of old and traditional agronomical practices. For example the Autoregressive Integrated moving average model (ARIMA) [21] and similar principles were used in predicting the crop yield based on soil condition, weather condition and cultivation practices [22]. Few major examples are discussed below.

2.1 GODAN Framework: Global Open Data for Agriculture and Nutrition

Open Data is a powerful, evidence-based tool for long- term sustainable development by improving economic opportunities for farmers and health of consumers.

Open access to research, meta-analysis, and open publication of data are vital resources for nutritional security. This tool delivers it.

GODAN can help achieve key characteristics such as,

- Open Data Policy: Transparency and congregate ownership
- Access to USDA: Vast kinds of data available to be extrapolated
- Data collaboration and Cooperation: Mutual Growth initiatives
- Minimizing Duplication: Helps in sorting out redundant data (Fig. 1).

2.2 Use of Drones in Agriculture

Huge Data innovation can be consolidated in Agriculture and united field by the utilization of automatons.

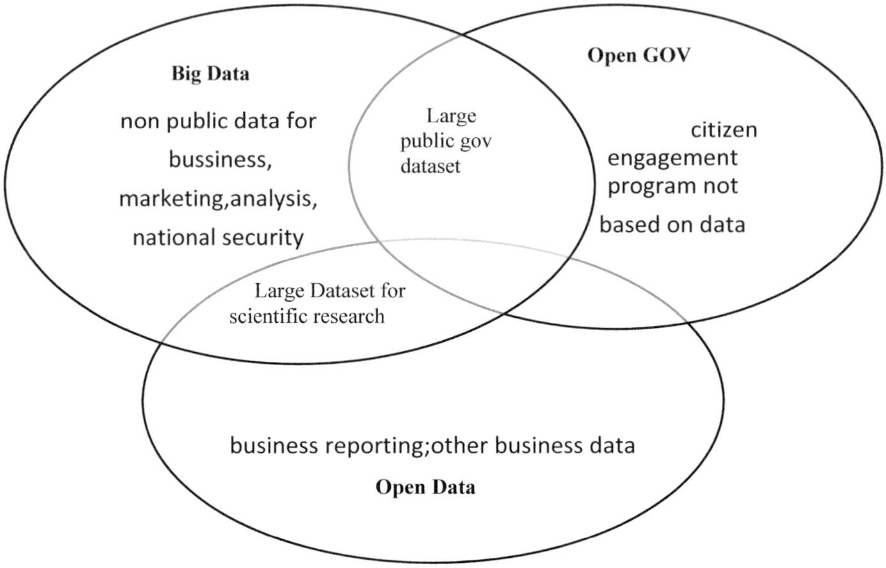

Fig. 1 GODAN framework: global open data for agriculture and nutrition. *Source* Wolfert et al. [4]

In a large portion of the created countries, utilization of brilliant machines and atmosphere sensors is developing to deal with the product. These brilliant machines and sensors likewise give cultivate information in enormous amount and which could be used to im-demonstrate cultivate related basic leadership process. Cultivating procedures will turn out to be progressively information driven and information empowered. Quick advancements in the Internet of Things and Cloud Computing are adding to the idea of Smart Farming [23]. While Precision Agriculture is simply taking in-field changeability into air conditioning tally, Smart Farming goes past that by putting together administration errands with respect to area as well as on information, improved by setting and circumstance mindfulness, activated by constant occasions. Ongoing helping reconfiguration highlights are required to do coordinated activities, particularly in instances of abruptly changed operational conditions or different conditions (e.g., climate, sickness, bother pervasion and so on). Drones could be utilized in joining outer Big Data sources, for example, climate or market information or benchmarks with different homesteads and along these lines making a comprehensive framework for productive ranch the board rehearses. Automatons in agribusiness and in savvy cultivating is extremely viable because of the way that unmanned aeronautical vehicles (UAV) can give ranchers a 10,000 foot perspective of their field's as yet staying near the landscape thus giving more exact assessments [24].

2.3 Create Better Supply Chain Optimizations

This objective can be achieved from study of secondary data on distribution system of agro foods and primary data coming from in-depth interviews of channel actors and industry experts. This objective can be used in developing the value chain mapping for agricultural foods throughout the world or even in any particular country or region (Fig. 2).

To come up with such a model various types of factors must be were analyzed using Confirmatory Factor Analysis followed by Linear Discriminate Analysis.

Confirmatory Factor Analysis: CFA is a multivariate statistical procedure that is used to test how well the measured variables represent the number of constructs. Researchers can specify the number of factors required in the data. It helps to identify that which measured variable is related to which latent variable. Confirmatory factor analysis (CFA) is a tool that is used to confirm or reject the measurement theory. This is done in the Data Analytics part.

Discriminate analysis is used when groups are known a priori (unlike in cluster analysis). Each case must have a score on one or more quantitative predictor measures and a score on a group measure. In simple terms, discriminate function analysis is classification—the act of distributing things into groups, classes or categories of the same type.

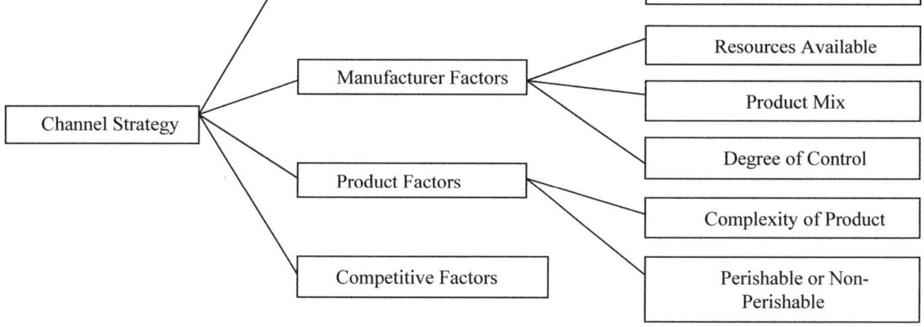

Fig. 2 Channel selection strategy for supply chain optimizations. *Source* Masih and Chauhan [28]

3 Using Online Data for Betterment of Farmers

3.1 A Case Study: Using Big Data Approach Enables Farmers to Buy Fertilizer from the Nearest Available Shop Thus Minimizing Transportation Cost

As indicated by World Bank's review, India has one of the biggest arable lands of about 159.7 million hectares. Agriculture forms the backbone of Indian economy. This calls for a high utilization of fertilizer in the country. Different fertilizer brands are available in the country. Getting fertilizer at affordable prices is top priority of the farmers. Transportation plays a major role in adding up to the cost of fertilizer. Lack of awareness and knowledge about the prices of fertilizers makes farmers helpless and they end up in purchasing costly fertilizers. We investigated this problem and tried to come up with a problem-solving Big Data approach. Our answer includes investigation of separations between spots where fertilizer is accessible and getting the farmer to the closest place to get the fertilizer. This diminishes the movement cost, accordingly lessening the cost required for purchasing the fertilizer.

In straightforward words, we can check whether there is a shop for a particular fertilizer in any region over a span of 50 km. The algorithm prepared would enable farmer to search over span of 50 km the closest region which could give adequate and number of fertilizer shops and details of merchant (either wholesalers or retailers, as determined by the client) in contrast with the closest locale. This algorithm will help farmers in saving their time, transportation cost and increase their barging power, since area that has more number of shops, offers more bargaining power to the farmer. The 50 km is set as the range on the grounds that further travel would expand the expense of the client (by and large which would be an agriculturist or farmer).

3.2 Methodology

We wish to compute the nearest districts so that farmer has to travel the least, for the purchase of fertilizer. However, we also wish to focus on real-time scenarios wherein we could return a district as output which is much closer to the user as well as have a good number of dealers available (as specified by the user: whether he/she wants to purchase the fertilizers in wholesale or retail). In case the user is searching for a district which has dealers (as specified) already present, we simply return the same district as the output.

In simple words, we check if there is a certain district in the 50-km radius of the nearest district which could provide sufficiently large number of dealers (either wholesalers or retailers, as specified by the user) in comparison with the nearest district. This shall allow the user to benefit in terms of availability of dealers at time of their visit as well as price of the fertilizers could be bargained more where there is high competition. The 50 km is set as the radius because further travel would increase the cost to the user (generally which would be a farmer) and thus, travel overhead would overcome the benefits reaped from bargaining the fertilizer prizes. In this way the model aims at providing the most fruitful deal for the farmers (Fig. 3).

The first value corresponds to the unique *ID* value assigned to each state in India. Second value corresponds to the name of the state. Third value corresponds to the name of the district in that particular state. The last two integer values represent the number of wholesalers and retailers, respectively, in the district.

The given dataset contains data for 598 districts all over India.

Further, we converted this data to ".JSON" format with additional properties. For each row we will find the latitude and longitude of the district specified. The latitude and longitude are found using the *chromedriver.exe* app. As we are using Python 3, we could make use of the *selenium* library which facilitates in the running the *webdriver* functions. The driver allows us to open a new Google Chrome tab. We put the google maps *URL* with the given district and state names as the *query* in the URL. The motivation behind using this approach is that the URL loaded onto the new window of the browser changes as the search is completely loaded. The new URL now contains the latitude and longitude value for the search query (city and state) specified.

The initial URL looks like: https://www.google.com/maps/place/City,State/

```
7,Chhattisgarh,Koriya,12,77
7,Chhattisgarh,Mahasamund,24,157
7,Chhattisgarh,Narayanpur,1,8
7,Chhattisgarh,Raigarh,24,202
7,Chhattisgarh,Raipur,88,494
7,Chhattisgarh,Rajnandgaon,28,210
7,Chhattisgarh,Surguja,43,179
7,Chhattisgarh,Uttar Bastar Kanker,28,133
8,Dadra & Nagar Haveli,Dadra & Nagar Haveli,0,1
9,Daman & Diu,Daman,0,1
10,Goa,North Goa,9,53
10,Goa,South Goa,9,43
11,Gujarat,Ahmadabad,65,418
11,Gujarat,Amreli,41,358
11,Gujarat,Anand,91,504
11,Gujarat,Banas Kantha,74,867
11,Gujarat,Bharuch,48,229
11,Gujarat,Bhavnagar,43,586
11,Gujarat,Dohad,27,195
11,Gujarat,Gandhinagar,42,272
11,Gujarat,Jamnagar,50,423
11,Gujarat,Junagadh,56,593
11,Gujarat,Kachchh,44,328
```

Fig. 3 The input data in the ".CSV" format (Comma-Separated Values format of dataset which includes unique ID of state, name of the state, name of the District, number of wholesalers and number of retailers in order). *Source* Author's self-computed data in Python

The city and state could be searched as needed by placing it in the above URL. The new updated URL looks like:
https://www.google.com/maps/place/Vellore,+Tamil+Nadu/@12.8992994,79.0483018,12z/data=!3m1!4b1!4m5!3m4!1s0x3bad38e61fa68ffb:0xbedda6917d262b5e!8m2!3d12.9165167!4d79.1324986

As we can see the text highlighted in the new URL is the co-ordinates of the place searched. The selenium library's *webdriver* function now allows us to fetch this co-ordinate. This allows us to find the latitude and longitude for each district already present in the dataset. Pre-processing and storing these values will allow us to directly compute the output for the user specified query.

These will be the 2 additional properties for the district. The processed data will be JSON file having a list of dictionary objects. Each dictionary object corresponds to one of the districts present in the dataset. We could visualize the JSON file as:

.... 453}, {"ID": 11, "City": "Rajkot", "State": "Gujarat", "Wholesalers": 85, "Retailers": 851, "Latitude": 22.2734269, "Longitude": 70.6812104}, {"ID": 1 ....

The set of all latitudes and longitudes of each district can be treated as a "*Network of Chemical Fertilizer Dealers*".

After the pre-processing phase, we are ready to process the user queries. The user can enter their choice of district and corresponding state (within India) from where they wish to start their travel. In any case, if user fails to correctly enter the name of the district or state, the driver tries to use its auto-correction ability while loading the initial URL. Yet, if the browser fails to search for the specified location, the driver displays the *default location* of the server which shall host the API. Correspondingly, the code checks if the co-ordinates retrieved from the *webdriver* are same as that of server's exact location, then it returns an error message asking the user to retry his/her search query.

If the query is successfully processed, we shall have now the co-ordinates of the location as specified by the user. More specifically, we can also allow the user to turn on their device GPS to automatically detect their location while we publish the application for public use.

Meanwhile, the received co-ordinates are then used to find distance with each of the co-ordinates of the districts as specified in the processed dataset. We use the *Haversine formula* to compute the distance between each pair of co-ordinates, correspondingly storing the distance value. The formula can be seen as:

$$\text{Let radius of earth be}(R) : 6371\text{ km}$$

Next, we write each latitude value as radians.

$$\varphi_1 = \frac{\text{Latitude}_1 \times \pi}{180}\text{radians}$$

$$\varphi_2 = \frac{\text{Latitude}_2 \times \pi}{180}\text{radians}$$

Followed by, we compute:

$$\Delta\varphi = \frac{(\text{Latitude}_1 - \text{Latitude}_2) \times \pi}{180} \text{radians}$$

$$\Delta\omega = \frac{\left(\text{Longitude}_1 - \text{Longitude}_2\right) \times \pi}{180} \text{radians}$$

$$a = \sin^2\left(\frac{\Delta\varphi}{2}\right) + \left[\cos(\varphi_1) \times \cos(\varphi_2) \times \sin^2\left(\frac{\Delta\omega}{2}\right)\right]$$

$$c = 2 \times a \tan 2\left(\frac{\sqrt{a}}{\sqrt{1-a}}\right)$$

Finally,

$$\text{The distance between 2 points is}(d) : R \times c \,\text{km}$$

This value is computed between the user-specified district and each of the districts in the given dataset. Then result is given as output using the approach as specified in the beginning. The user is able to see the place of travel, the number of wholesalers or retailers available (as mentioned by him/her) and the distance to be travelled (Fig. 4).

The above-mentioned example of self-prepared and tested case study is aimed at benefitting farmers to the maximum who do not have proper contacts or whose contact do not have proper affiliation with the Agricultural division and are being deprived of good fertilizers and seeds. Through this algorithm, farmers could be made aware of the nearest Government certified Fertilizer center's with proper latitude and longitude and they can buy proper fertilizers for their lands. The project yields the nearest fertilizer center or shop in a 50-km radius of their location. Moreover, our project provides options of searching for Wholesale or Retail shops. The project has potential of boosting agriculture productivity by helping farmers to get access to the '**right fertilizer at right time from right shop at right price**'.

```
Enter the name of your city or district: Bhuj

Enter the name of the state: Gujarat

1. Wholesalers OR 2. Retailers: 2

Destination: Jamnagar, Gujarat
Distance: 95.11 km
Retailers: 423
```

Fig. 4 Query regarding a district named Bhuj in the state of Gujarat. *Source* Author's self-computed data in Python

For future scope, the project can be established as an app, so that farmers can access the data anywhere, anytime hence featuring concept of ubiquitous computing. It will benefit farmers to a very large extent.

4 Use of Social Media Analytics in Agriculture

Sentiment analysis is the study of moods, opinions, emotions, evaluations and attitudes toward a specific subject or field or topic [25]. Sentiment analysis is a type of natural language processing for tracking the mood of the public about a particular product or topic. Sentiment analysis helps in classification of opinion as positive, negative or neutral [26]. Sentiment analysis can be useful in several ways. For example, in marketing it helps in judging the success of an ad campaign or new product launch, and even identify which demographics like or dislike particular features [27]. The difference between sentiment analysis from editorial media and social media (Facebook, Twitter, Youtube) is that in social media text is short, usually one sentence composed of at most 140 characters whereas editorial data is long editorial piece of information which includes online books, journals, research papers, magazines, where relevant information is pointed out as hit sentence or match sentence [28, 29]. Sentiment analysis in agriculture sector is gaining momentum around the world to deeply understand the issues related to crop production, dairy and meat products including specialty foods like gluten-free foods, dairy-free food, halal meat, etc. [30]. In this section of the chapter, we have explained how sentiment analysis can be used in understanding the leading trends, opinions and consumer preferences in agriculture taking the example of gluten-free foods. Globally gluten-free products are estimated to have a market potential of 7594.43 Million USD in 2020. Reasons for purchase of gluten-free foods are: 25% consumers considered them healthy, 19% consumed them for weight management, 18% considered them trendy, 17% considered them of higher quality than normal food products, 16% purchased them due to less carbohydrates and 5% due to celiac disease or wheat allergy. Here using data from editorial media and social media, public sentiment was studied. Data was taken for year 2017–18. Generally data for social media and editorial media can be obtained by two ways:

(a) **Free data extraction tools**: It can be obtained from google trends (see Example https://trends.google.com/trends/explore?q=gluten-free), other site like Twazzup, Boardreader, HowSociable, Likealyzer and Social Mention allows a user to get the social media data up to certain limit for free. They also provide information like top tweets or posts, top influencers or followers, most trending news, engagement rate, etc. which could be used to enhance your research.

(b) **Paid data extraction tools**: Data mining companies like Meltwater, Sysomos and Rival IQ provide data in bulk quantities and it can be used in research. These companies also provide additional services like sentiment analysis, heap maps, top posts, word cloud, etc.

In addition sentiment analysis could be done in two ways:

(a) **Manual sentiment analysis**: Here after the data extract in excel file, the researcher goes through all the social media or editorial media content by himself and rate the sentiments as positive, negative or neutral, depending upon the relevance of the content. Manual sentiment analysis could be done for qualitative researches, where every word holds a specific relevance. It is generally used in researches related to consumer behavior, competitor's analysis and SWOT analysis (Strength-weakness-opportunity-threat).
(b) **Machine learning based sentiment analysis**: There are many online sites where you can get the sentiment analysis done for the data. Examples of few such sites are Hootsuite Insights, Quick Search, NCSU Tweet Visualizer, MeaningCloud, Sentiment Analyzer, SentiStrength and Sentigem.

In our example we have used the data from Perspectory Media Insights (a paid site for social media and editorial media data), since paid sites provide data volume in bulk and makes the research more efficient. However, the beginners can also use free data site for exploration purpose. Some data was also obtained from Google trends. Machine learning-based sentiment analysis was used in research to get the trends related to gluten-free foods. However, manual sentiment analysis was also performed as a cross check to Machine learning-based sentiment analysis and was found up to 95% accurate.

4.1 *Understanding the Monthly Global Trends of Gluten-Free Foods*

Figure 5 shows monthly trend of gluten-free foods in year 2017–18. Data volume of social media is very high compared to editorial media. People post many things on social media especially on Twitter about gluten-free diet. However, in editorial

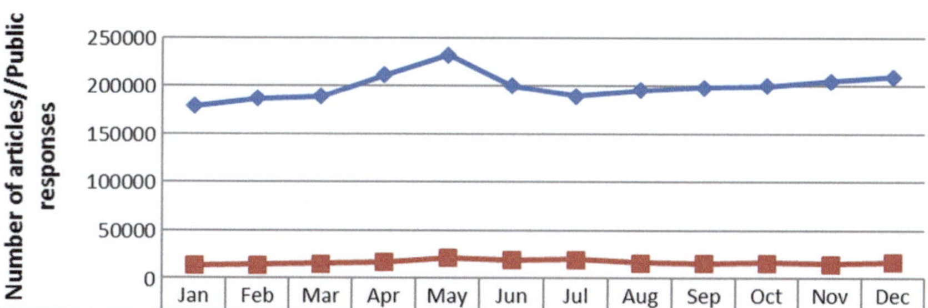

	Jan	Feb	Mar	Apr	May	Jun	Jul	Aug	Sep	Oct	Nov	Dec
Social Media	2E+05	2E+05	2E+05	2E+05	2E+05	2E+05	2E+05	2E+05	2E+05	2E+05	2E+05	2E+05
Editorial Media	13530	14250	15555	16872	21389	19060	19741	16397	16166	16556	15178	17072

Fig. 5 Monthly global trends for editorial and social media 2017–18. *Source* Author's own compilation of editorial and social media data obtained from Perspectory Media Insights

	A	B	C	D	E	F	G	H
1	Date	Headline	URL	Hit Sentence	Source	Influencer	Country	Sentiment
2	21-Jan-2017 09:08AM	West Ham	http://www	or lunch and monthly supperclub. Round off with a Bake-a-Boo gluten-free cupcake.\n\nThe Kitchen Table Mill Lane\n\nLa Petite Corée is a local	Hampstea	Prudence I	United Kin	Neutral
3	21-Jan-2017 09:08AM	Pass the d	http://www	sales, is offering that product — nuggets of dogfish coated in a gluten-free, allergen-friendly crust. Bartholomew, who believes so-called	Mysask.co	-	Canada	Neutral
4	21-Jan-2017 09:08AM	Privacy Im	http://www	party for 12 guests. My friends had a variety of dietary needs: vegetarian, gluten-free, dairy-free and tomato-free. I had considered these	SCL	-	United Kin	Neutral
5	21-Jan-2017 09:08AM	Artificial S	https://sg.	reports based on specific client requirement.\n\nRelated reports\n\n- Global Gluten-Free Packaged Food Market 2015-2019- Frozen Food Packaging	Yahoo! Sin	Reportlink	Singapore	Neutral
6	21-Jan-2017 09:08AM	Top nine f	http://www	more, less or none of any ingredient, and it has introduced gluten-free dishes.\n\nThe piccolo menu also offers several types of pasta with	Get Readi	By Frances	United Kin	Neutral
7	21-Jan-2017 09:08AM	Global Flo	https://sg.	. The growing demand for bread and bakery products and the awareness regarding gluten free food items are driving the growth in this region.	Yahoo! Sin	Reportlink	Singapore	Neutral
8	21-Jan-2017 09:08AM	Nora and I	http://blog	, what would it be? Is it cake?\n\nNora: Chocolate!\nLaura: Nora's gluten-free goat cheesecake. It's just the best!\n\nAw! We all wish we could	Bloesem	Irene Hoof	Singapore	Positive
9	21-Jan-2017 09:08AM	Smoked ha	http://www	eat! Perfect for breakfast, brunch or even dinner, it's also gluten-free, low-carb and packed with calcium and Vitamin D for healthy bones!	Cook Siste	-	United Kin	Positive
10	21-Jan-2017 09:08AM	Blaze Fast	http://be.s	. For pizza fans with specific dietary needs, Blaze Pizza offers gluten-free dough and vegan cheese. The new Knoxville restaurant will also	SYS-CON E	Marketwir	Belgium	Positive
11	21-Jan-2017 09:08AM	Blaze Fast	http://de.s	. For pizza fans with specific dietary needs, Blaze Pizza offers gluten-free dough and vegan cheese. The new Knoxville restaurant will also	SYS-CON C	Marketwir	Germany	Positive
12	21-Jan-2017 09:08AM	Knorr laun	http://www	which, for the first time, communicates on pack that the cubes are gluten-free, as well as bringing to life the core values of the brand and	Convenien	-	United Kin	Positive
13	21-Jan-2017 09:08AM	Legacy Ve	http://www	want to put into your body and nothing you don't. They are non-GMO, gluten-free, paleo-friendly, vegan, raw and peanut-free.\n\nRon Patel,	Stockhous	LGYV	Canada	Neutral
14	21-Jan-2017 09:08AM	Attention I	http://www	superior performance but a new study suggests that the benefits of a gluten free diet are greatly exaggerated.\n\nMore than 1000 athletes took	Hindustan	-	India	Neutral

Fig. 6 Example of raw data extracted from social and editorial media. *Source* Author's own compilation of editorial and social media data obtained from Perspectory Media Insights

media only journalists, blog writers and researches publish their findings and opinions about gluten-free food. April and May had highest number of editorial articles and social media public responses as month of May was celebrated as "Celiac Awareness Month". Lot of dieticians and blog writers wrote about gluten-free foods and related benefits. After May, the volume of articles reduced but increased again in December due to Christmas and New Year week. On social media people posted about gluten-free snacks, desserts, new recipes. Many restaurants and hotels updated their menu with new recipes of gluten-free foods (Fig. 6).

Figure 7 tells about most searched or used keywords on editorial and social media. It also tells about growth percentage of these keywords in year 2017–18. McDonalds gluten-free menu was most searched topic (growth rate of 650%), since it is affordable and tasty. Brown sugar, gluten-free pancake recipe butter cookie, fish and chips and Rice Krispies Treats reported growth rate above 150% due to availability new food variants and good taste.

4.2 Global Sentiment Analysis for Gluten-Free Foods

Figure 8 shows sentiment analysis via editorial media. The trends report that leading positive sentiments (53%) related to gluten-free foods talked about organic food, marshmallows, reduce weight, oatmeal, Cheerios, Arrowroot, Peanut butter cookie. Neutral sentiments (43%) talked about cookies, allergy, beer, bakery, bread, bun,

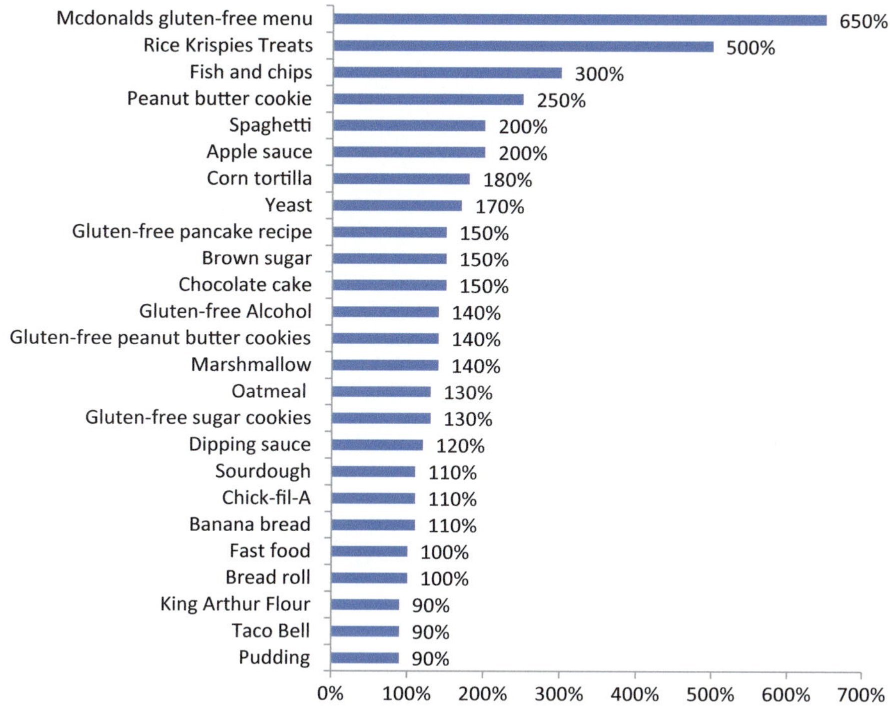

Fig. 7 Growth percentage of popular keywords on editorial and social media for gluten-free foods 2017–18. *Source* Author's own compilation from google trends, editorial media data and social media data from Perspectory Media Insights

Fig. 8 Global sentiment analysis for editorial media 2017–18. *Source* Author's own compilation from editorial media data from Perspectory Media Insights

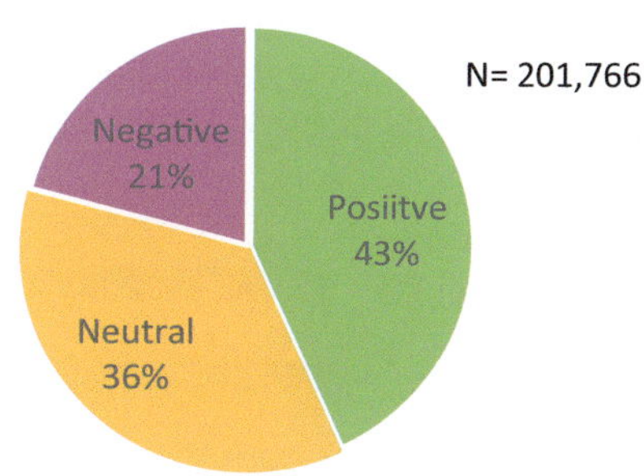

recipes, cracker, Negative sentiments (4%) considered gluten-free foods as a fad (short-term dietary trends), less nutritious which lacks dietary fiber, lacks essential nutrients and is very costly compared to other conventional foods.

Figure 9 shows sentiment analysis via social media. On social media, positive sentiments (32%) talked about good taste of gluten-free rice Krispies, bakery products, dark chocolates and gluten-free oats. Neutral sentiments (51%) discussed about availability of gluten-free, dairy-free and vegan food options, along with availability

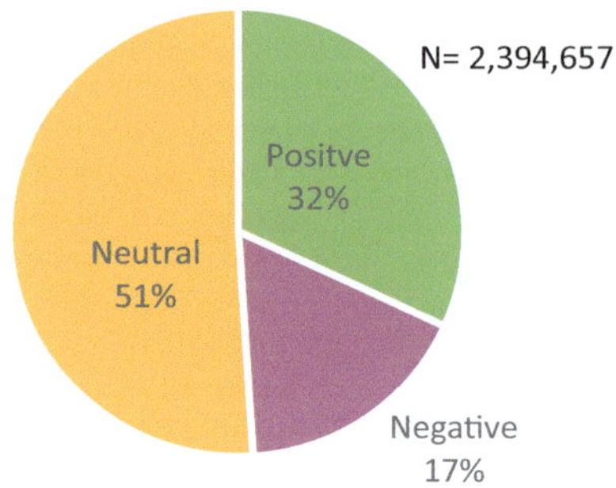

Fig. 9 Global sentiment analysis for social media 2017–18. *Source* Author's own compilation from social media data from Perspectory Media Insights

of gluten-free protein bar, pop-corn, and Indian grains in the supermarkets and specialty stores. Negative sentiments (17%) discussed about gluten-free foods as cause of obesity in lots of people since they have high sugar and carbohydrate content. Some people also complaint about dry taste of gluten-free bread and cookies, others complaint about lack of availability of desired gluten-free products in nearby store.

Figure 10 shows social analysis by channel. Mainly five channels were used to extract the data from social media namely Twitter, Facebook, Instagram, Blogs and Youtube.

Twitter (49.10% responses) micro blogs were written by food bloggers, dieticians, nutritionists and food experts on new gluten-free recipes, healthy gluten-free diet and easy to make gluten-free foods. On **Facebook** (23.16% responses) fan pages like Free-From-Gluten, Gluten Dude were created to provide information about availability of gluten-free foods, gluten-free facebook groups were formed to discuss more about celiac disease and food options available. At **Instagram** (14.52% responses) famous pages were #glutenfree, gluten.free.me. People posted pictures and share recipes of gluten-free foods. Many food bloggers wrote **Blogs** (12.22% responses) benefits and harms of consuming gluten-free diet. On **Youtube** (1% response) many videos are available to cook healthy gluten-free foods.

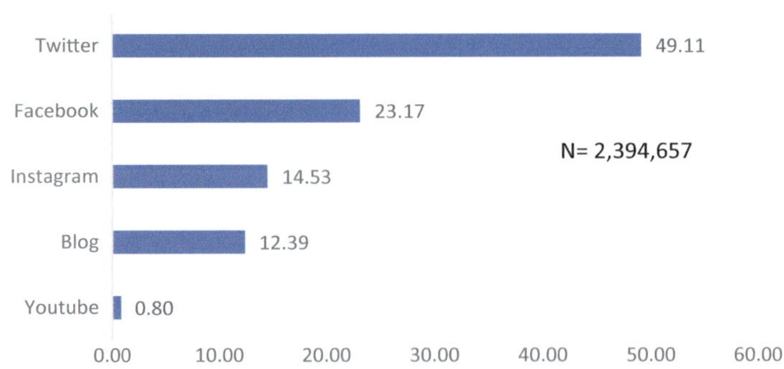

Fig. 10 Social media analysis by channel 2017–18. *Source* Author's own compilation social media data from Perspectory Media Insights

4.3 Perceptual Mapping Based on Sentiment Study

Perceptual mapping technique was used in the study to draw the trends from editorial media. Food categories taken into consideration were Flour and Mixes, fast food (including pizza and pasta), Bakery, Cereals and Cornflakes, Snacks, Soft and energy drinks, Liquor, Confectionary and last category was Desserts [28].

$$\begin{aligned}&\textbf{Score calculation for each food category} = (\textbf{No. of} +\textbf{ve article} * \textbf{1.5})\\&\quad + (\textbf{No. of neutral article} * \textbf{1}) + (\textbf{No. of} -\textbf{ve article} * \textbf{0.5})\\&\%\ \textbf{Share for each food category}\\&\quad = \frac{\textbf{Total Number of articles for a particular food category} \times \textbf{100}}{\textbf{Total Number of articles for all the food Categories}}\end{aligned}$$

Figure 11 shows a global perceptual map for different food categories of gluten-free foods based on the data of editorial media. Gluten-free flours and mixes were in maximum demand all over the world followed by pizza, pasta and other bakery items.

4.4 Use of Online Portals for Data Representation

Information from online portals can be used to enhance to research results. In this section, free data from portal of Google trends was used to create a heap map, thus representing the areas of world where maximum and minimum online researches of gluten-free foods were made.

In Fig. 12, dark blue regions denote the areas where maximum searches had been made about gluten-free foods. Medium blue color denote the areas where medium level of searches were made. Light blue color denote the areas where only few searches were made about gluten-free foods.

Gluten-free foods across the globe.

Figure 12 and Table 2 show the world scenario about the searches made on gluten-free foods. Dark blue color denoted the countries having maximum percentile between 70 and 100. It included 3 countries USA, Canada and UK. These countries had maximum access to gluten-free foods, as they were developed countries and had high availability of gluten-free foods with a lot of food variants available (Google trends, 2017) (Table 3).

Medium blue color denoted the countries having percentile between 50 and <70 including Australia, Mexico and Brazil. These 3 countries have good number of patients for celiac or wheat allergy (Google trends, 2017).

Light blue color denoted countries having percentile of 20 to <50. These were mainly developing countries like India, Morocco and many European countries. In

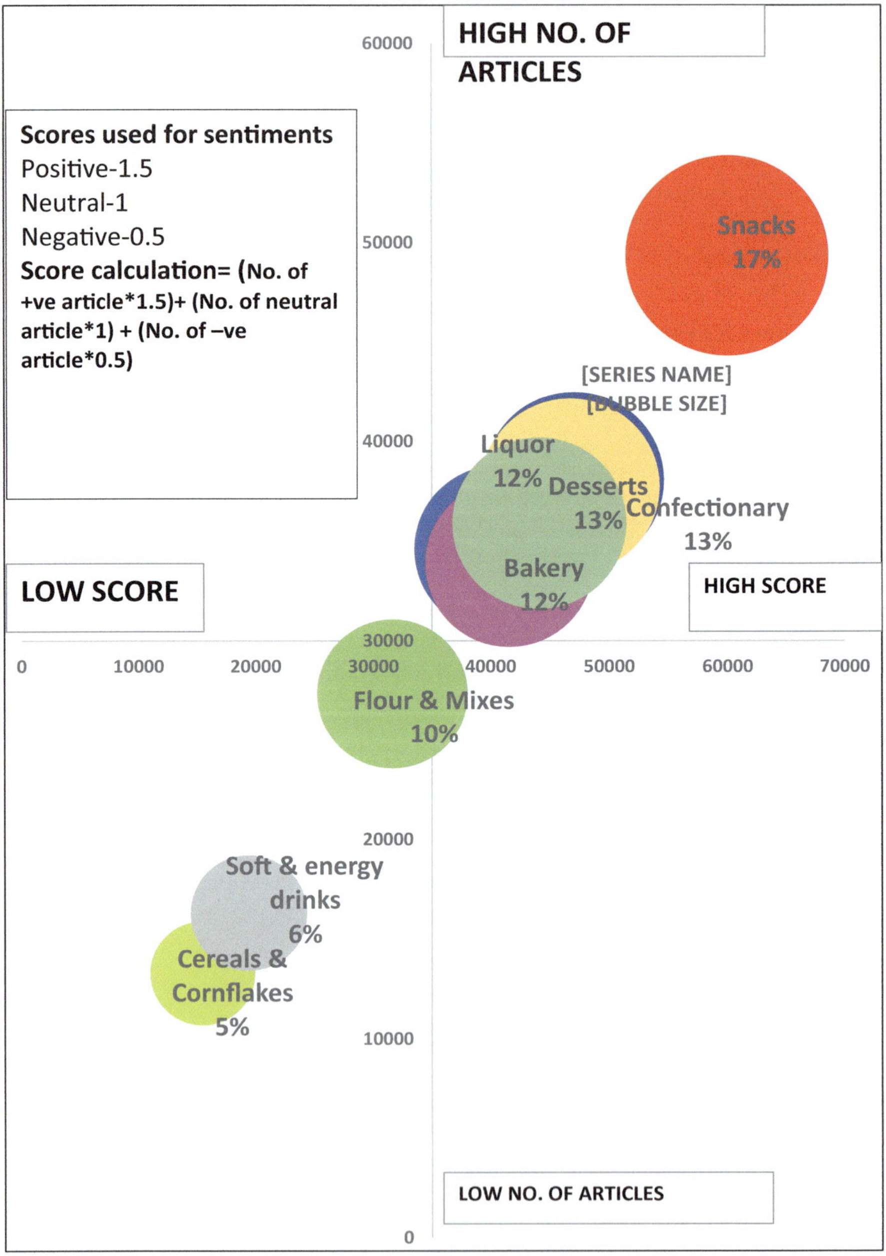

Fig. 11 Perceptual map of editorial media for different food categories of gluten-free food across the globe 2017–18

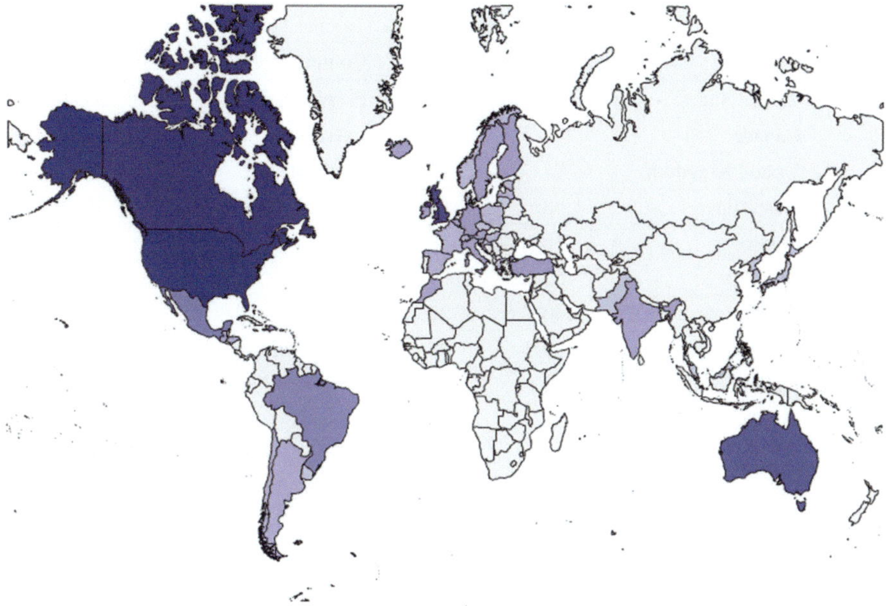

Fig. 12 Heat map for online searches on gluten-free foods-2017–18. *Source* Author's own compilation from google trends

Table 2 Score card

Food items	Score	No. of articles	% Share of articles (%)
Flour and mixes	31,626	27,328	10
Pizza, pasta, fast food	47,043	38,054	13
Bakery	40,718	34,585	12
Cereals and cornflakes	15,454	13,299	5
Snacks	60,031	49,336	17
Soft and energy drinks	19,375	16,324	6
Liquor	41,558	33,864	12
Confectionary	46,714	37,744	13
Desserts	44,085	35,891	13

Source Author's own compilation from editorial media data from Perspectory Media Insights

Table 3 Percentile of top 50 countries who made online searches about gluten

Rank	Country	Percentile	Rank	Country	Percentile
1	United States	100	26	France	26
2	Canada	79	27	Latvia	26
3	United Kingdom	74	28	Switzerland	26
4	Australia	68	29	Norway	25
5	Mexico	58	30	Poland	24
6	Brazil	52	31	Singapore	24
7	Turkey	40	32	Estonia	23
8	Germany	39	33	The Netherlands	23
9	Iceland	39	34	Bosnia and Herzegovina	22
10	Finland	38	35	Greece	22
11	Ireland	38	36	San Marino	22
12	Morocco	37	37	Vatican City	22
13	India	36	38	Uruguay	22
14	Sweden	36	39	Denmark	21
15	Guatemala	36	40	Slovenia	21
16	Dominican Republic	36	41	Pakistan	20
17	Austria	35	42	Croatia	20
18	Italy	35	43	Czech Republic	19
19	Honduras	33	44	Liechenstein	19
20	Belgium	32	45	Japan	19
21	Argentina	32	46	Hong Kong	19
22	Andorra	29	47	Lithuania	18
23	Spain	29	48	Luxembourg	18
24	Chile	29	49	Malaysia	18
25	Hungary	27	50	Malta	17

Source Author's own compilation from google trends

developing countries, gluten-free foods were at a very niche stage, and users here searched for very basic food options (Google trends, 2017).

4.5 To Study Consumer Behavior of Domestic and Processed Gluten-Free Food Using Discriminant Analysis

In this objective researcher created a Fan page of gluten-free foods on Facebook and requested users of gluten-free foods to fill in few details. Based to data collected from Gluten-free food Fan page, the feedback model using discriminant analysis was created. In this small social media survey analysis, researcher aimed to find out that which category of gluten-free foods, namely domestic, semi-processed and completely processed is most famous among the users.

This objective was analyzed using discriminant analysis model for consumers of gluten-free foods. The discriminant function score for the ith function is: $D_i = d_{i1}X_1 + d_{i2}X_2 + \ldots + d_{ip}X_p$. Where X = the score on each predictor, and d_i = discriminant function coefficient.

Dependent Variable categories: Three dependent variable categories are selected for the study: domestic food (completely domestic), semi-processed food (regularly domestic and irregularly processed), completely processed food (regularly processed and irregularly domestic).

Independent Variables: Z_1 = Age (years), Z_2 = gender (Male/female), Z_3 = Height (cm), Z_4 = Weight (kg), Z_5 = Region (rural/urban), Z_6 = BMI ratio, Z_7 = Bread/chapattis/day (number), Z_8 = Education, Z_9 = Occupation, Z_{10} = Monthly household income (USD), Z_{11} = Family Type (nuclear/joint), Z_{12} = Marital status (single/married/divorced/remarried), Z_{13} = Severity of disease (not allergic and consuming for health benefits/mild intolerance/moderate intolerance/severe intolerance/Hyper-insensitivity), Z_{14} = Type of diet (vegetarian/non-vegetarian/flexitarian), Z_{15} = Type of eater (whole foodie/British classic/fast foodie/super snacker/herbivore/can't say), Z_{16} = Frequency to visit doctor, Z_{17} = Implementation of health tips (yes/no), Z_{18} = Frequency of weekly exercise, Z_{19} = Average sleeping hours, Z_{20} = Daily working hours, Z_{21} = Bar and restaurant keep GF menu (yes/no), Z_{22} = Processed GF food at work place (yes/no), Z_{23} = Domestic GF foods offer higher customization of products (1 as completely disagree to 5 as completely agree scale), Z_{24} = Domestic GF foods has higher chances of contamination (1 as completely disagree to 5 as completely agree scale), Z_{25} = Most effective recommendation for GF food, Z_{26} = Biopsy, Z_{27} = Supplements, Z_{28} = Fruits and vegetables/month, Z_{29} = Bakery products/month, Z_{30} = Pizza or pasta/month, Z_{31} = Cornflakes or muesli/month, Z_{32} = Flour mixes/month, Z_{33} = Snacks/month, Z_{34} = Cold or health drinks/month, Z_{35} = Alcohol/month, Z_{36} = Confectionary/month, Z_{37} = Desserts/month.

An investigative online study was carried out to uncover the different types of food habits and dependencies of other variables on these food habits. The results for discriminant analysis are discussed below.

4.6 *Discriminant Analysis Model*

Model consists of 550 respondents. Dependent variables considered for combined discriminant analysis model are domestic foods (group 1), semi-processed food (group 2) and completely processed food (group 3) (Table 4).

Accuracy: 0.9166; 95% CI: (0.8898, 0.938); No Information Rate: 0.714, P-value 0.001.

Table 4 Confusion matrix and statistics-combined model

	Domestic GF food	Semi-processed GF food	Completely processed GF food
Domestic GF food	40	20	
Semi-processed GF food	22	368	
Completely processed GF food		4	95

Source Author's own compiled primary data from Facebook Fan Page of Gluten-free foods and computed by *R* programming

Confusion matrix tells about the number of cases correctly and incorrectly assigned to each of the groups based on the discriminant analysis. It compares predicted versus actual group membership. Confusion matrix is used to measure wellness of derived functions in predicting group membership [31] (Fig. 13).

We can identify the nature of the discrimination for each discriminant function by looking at the means for the functions across groups. Group means are centroids. Differences in location of centroids show dimensions along which groups differ. Functions discriminate between groups by plotting the individual scores for the two

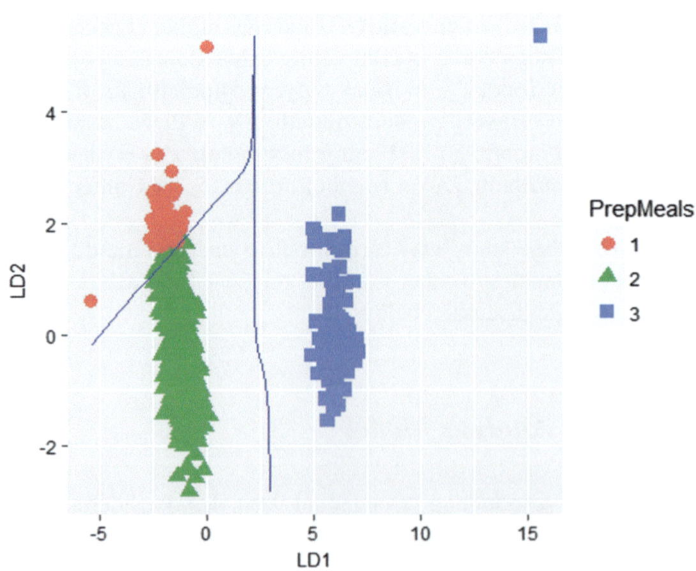

Fig. 13 Centroid mapping for combined model. *Source* Author's own compiled primary data from Facebook Fan Page of Gluten-free foods and computed by *R* programming

discriminant functions. As visible in the graph centroids for group 1 (domestic food) are in red color. Centroids of group 2 (semi-processed category) are in green color and are maximum in number. Group 2 is most prominent and strongest group. Centroids of group 3 (completely processed food) are in blue color and are quite separated from group 1 and 2. However, group 1 and 2 seems to overlap each other in few cases.

It was concluded through discriminant analysis that most the customer like semi-processed gluten-free since they can prepared them according to their own taste and health requirements. Completely processed gluten-free was second most preferred food category, it was more popular in students and working women who had limited time for cooking. Completely domestic gluten-free foods were least preferred since it was time-consuming and had higher chances of contamination with other wheat-based products.

The discriminant function coefficients denote the unique contribution of each variable to the discriminant function.

$D_1 = -0.005833010\, X_1 + 0.007028640\, X_2 + 0.7180335\, X_3 + 0.01716554\, X_4 + 0\, X_5 + 0.01716554\, X_6 + 0.05\, X_7 + 0.04\, X_8 - 0.05338409\, X_9 + 0.00\, X_{10} + 0.19\, X_{11} + 0.06\, X_{12} + 0.04778012\, X_{13} - 0.07122121\, X_{14} + 0.04\, X_{15} + 0.03\, X_{16} - 0.2825933\, X_{17} + 0.01\, X_{18} - 0.07924684\, X_{19} - 0.2390668\, X_{20} - 2.45\, X_{21} - 3.571549\, X_{22} - 0.14\, X_{23} + 0.14\, X_{24} + 0.05204288\, X_{25} - 0.08\, X_{26} + 0.05698819\, X_{27} - 0.04\, X_{28} + 0.02\, X_{29} - 0.2540003\, X30 + 0.02\, X31 - 0.1330998\, X_{32} + 0.02\, X_{33} + 0.04\, X_{34} - 0.04407397\, X_{35} + 0.01\, X_{36} + 0.02\, X_{37}.$

$D_2 = 0.01\, X_1 - 0.1684566\, X_2 + 0.7480926\, X_3 + 0.05200148\, X_4 + 0.00\, X_5 + 0.05200148\, X_6 + 0.07\, X7 - 0.02\, X_8 + 0.07591531\, X9 - 0.0000008428384\, X_{10} + 0.35\, X_{11} - 0.04283165\, X_{12} + 0.00\, X_{13} - 0.06946207\, X_{14} + 0.03\, X_{15} + 0.10\, X_{16} + 0.36\, X_{17} + 0.17\, X_{18} + 0.02\, X_{19} - 0.3233144\, X_{20} + 0.30\, X_{21} - 1.28\, X_{22} + 0.03\, X_{23} - 0.35\, X_{24} - 0.09\, X_{25} - 0.13\, X_{26} + 0.1150664\, X_{27} - 0.10\, X_{28} - 0.2540003\, X_{29} + 0.01\, X_{30} - 0.04\, X_{31} - 0.2014740\, X_{32} - 0.09\, X_{33} + 0.04\, X_{34} + 0.10\, X35 - 0.0003397069\, X_{36} - 0.03\, X_{37}.$

These discriminant scores are predictors of group membership for dependent variables [32].

References

1. Li, X., Chen, S., Guo, L.: Technological innovation of agricultural information service in the age of Big Data. J. Agric. Sci. Technol. **6**(4), 1008–0864 (2014)
2. Devlin, B.: The Big Data Zoo—Taming the Beasts: The Need for an Integrated Platform for Enterprise Information. 9sight Consulting, Cape Town (2012)
3. UNECE Classification of types of Big Data. Retrieved from http://www1.unece.org/stat/platform/display/bigdata/Classification+of+Types+of+Big+Data. (2013). Accessed 12 May 2018
4. Wolfert, S., Ge, L., Verdouw, C., Bogaardt, M.J.: Big data in smart farming—a review. Agric. Syst. **153**, 69–80 (2017)
5. Anonymous: Technology helps farmers to cater climate changes effects. Flare. Retrieved from http://www.flare.pk (2014). Accessed 12 May 2018

6. Grobart, S.: Dairy industry in era of Big Data: new gadgets help farmers monitor cows and analyze their milk (2012)
7. Vogt, W.: Looking at Big Data One Plant at a Time. Farm Industry News (2013)
8. Sonka, S.: Big Data: from hype to agricultural tool. Farm Policy J. **12**(1), 1–9 (2015)
9. Royse, R.: The Growing Field of Agriculture Technology. Vator News (2014)
10. Lane, J.: Digital soil: the four secrets of the new agriculture. Biofuels Digest (2015)
11. Guild, M., Danaher, T.: Big data comes to the farm. Financial Sense (2014)
12. Kwasi, B.R.: Long run and short run causality of rice consumption by urbanization and income growth in Ghana. ACADEMICIA: Int. Multi. Res. J. **5**(2), 173–189 (2015)
13. Mishra, P., Sahu, P.K., Dhekale, B.S., Vishwajith, K.P.: Modeling and forecasting of wheat in India and their yield sustainability. Indian J. Econ. Dev. **11**(3), 637–647 (2015)
14. Kwasi, B.R., Kobina, B.J.: Forecasting of cassava prices in the central region of Ghana using Arima model. Intercontinental J. Market. Res. Rev. 2(8) (2014a)
15. Kwasi, B.R., Kobina, B.J.: Cassava markets integration analysis in the central region of Ghana. Indian J. Econ. Dev. **10**(4), 319–329 (2014b)
16. Burark, S.S., Pant, D.C., Sharma H., Bheel, S.: Price forecast of Coriander—a case study of the Kota Market of Rajasthan. Indian J. Agric. Market. **27**(3), 72 (2013)
17. Harvey, A.C., Todd, E.C.: Forecasting econometric time series with structural Box-Jenkins models (with discussion). J. Bus. Econ. Stat. **1**(4), 299–315 (1983). Retrieved from http://www.jstor.org/pss/1391661. Accessed 12 May 2018
18. Juselius, K.: The Co-integrated VAR Model: Methodology and Applications. Oxford University Press, New York (2006)
19. Mafimisebi, T.E., Agunbiade, B.O., Mafimisebi, O.E.: Price variability, co-integration and exogeneity in the market for locally produced rice: a case study of southwest zone of Nigeria. Retrieved from http://dx.doi.org/10.4172/jrr.1000118 (2014). Accessed 12 May 2018
20. Paul, R.K.: Forecasting wholesale price of pigeon pea using long memory time-series models. Agric. Econ. Res. Rev. **27**(2), 167–176 (2014)
21. Bhardwaj, S.P.: Significance of market information system (MIS) in agricultural development. Indian J. Agric. Market. **25**(3), 83–84 (2011)
22. Brockwell, P., Davis, R.: Introduction to Time Series and Forecasting. Springer, New York (2002)
23. Sundmaeker, H., Verdouw, C., Wolfert, S., Pérez Freire, L.: Internet of food and farm 2020. In: Vermesan, O., Friess, P. (eds.) Digitising the Industry-Internet of Things Connecting Physical, Digital and Virtual Worlds, pp. 129–151 (2016)
24. Tripicchio, P., Satler, M., Dabisias, G., Ruffaldi, E., Avizzano, C.A.: Towards smart farming and sustainable agriculture with drones. In: 2015 International Conference on Intelligent Environments (IE), pp. 140–143. IEEE (2015)
25. Liu, B.: Sentiment analysis and opinion mining. Synth Lect. Hum. Lang. Technol. **5**(1), 1–167 (2012)
26. Denecke, K., Nejdi, W.: How valuable is medical social media data? Content analysis of the medical web. Inf. Sci. **179**, 1870–1880 (2009)
27. Qiu, G., He, X., Zhang, F., Shi, Y., Bu, J., Chen, C.: DASA: dissatisfaction-oriented advertising based on sentiment analysis. Expert Syst. Appl. **37**, 6182–6191 (2010)
28. Masih, J., Chauhan, S.: Impact assessment of Nielsen as a research firm using editorial online media. Br. J. Market. Stud. **3**(3), 45–55 (2015)
29. Montoyo, A., MartíNez-Barco, P., Balahur, A.: Subjectivity and sentiment analysis: an overview of the current state of the area and envisaged developments (2012)
30. Mostafa, H.: Supervised learning based on temporal coding in spiking neural networks. IEEE Trans. Neural Netw. Learn. Syst. **29**(7), 3227–3235 (2018)
31. Kwon, O.W., Chan, K., Hao, J., Lee, T.W.: Emotion recognition by speech signals. In:Eighth European Conference on Speech Communication and Technology (2003)
32. Loewe, V., Navarro-Cerrillo, R.M., García-Olmo, J., Riccioli, C., Sánchez-Cuesta, R.: Discriminant analysis of Mediterranean pine nuts (Pinuspinea L.) from Chilean plantations by near infrared spectroscopy (NIRS). Food Control **73**, 634–643 (2017)

Solar-Powered Automated IoT-Based Drip Irrigation System

Ananya Barman, Biswarup Neogi and Souvik Pal

Abstract In India the main economic structure is based on the agriculture sector. Worldwide, India is in 2nd position regarding farm output. Irrigation is an important component of agriculture. In most of the cases, conventional electricity or diesel-generated electricity is used by the farmers to run the motors and pumps manually, resulting in over irrigation and wastage of water. The use of conventional energy resources like petrol, diesel, coal is expensive and also creates air pollution and noise. Hence nowadays maximum research works are focusing on utilization of renewable energy resources, which is widely available, like solar energy. Solar-powered irrigation system is not only eco friendly but also economically a cheap process. This process minimizes the operation and maintenance cost, which is 2–4 times higher than a solar photovoltaic (PV) pump. In most of the developing countries, agriculture is solely dependent on rain water and thus poorly affected in the season of summer. In India there are also some of the places where water availability is very low which has detrimental effects on crop production. Hence nowadays wireless sensors were used to analyze the moisture level of the field and to automate the irrigation process. An automated irrigation system refers a system, with a minimum of manual requirement. All types of irrigation process like drip irrigation, sprinkler irrigation or surface irrigation can be automated by using timers, sensors or computers or mechanical appliances. In this automated system, some standard data regarding the moisture level of a field is predefined for a specific crop. The sensor first transmits the data through a microcontroller to the server after a specific time interval. If

A. Barman (✉)
Department of Chemistry, JIS College of Engineering, Kalyani, India
e-mail: ananya.barman@jiscollege.ac.in

A. Barman · B. Neogi
Department of Electronics and Communication Engineering, JIS College of Engineering, Kalyani, India
e-mail: biswarup.neogi@jiscollege.ac.in

A. Barman · S. Pal
Department of Computer Science and Engineering, JIS College of Engineering, Kalyani, India
e-mail: souvikpal22@gmail.com

P. K. Pattnaik et al. (eds.), *IoT and Analytics for Agriculture*, Studies in Big Data 63,
https://doi.org/10.1007/978-981-13-9177-4_2

the amount of moisture of the field is below the standard value of the database, the microcontroller will trigger the pump for watering the plants till the predefined moisture level. While the moisture level of the field reaches the threshold limit, the pump will automatically stop watering through microcontroller. The sensor analyzes the moisture level by monitoring the field in a specific time slot and store the data in database. Nowadays there are so many advancements taking place in sensing and communication technologies. Thus in today's life solar-powered automated drip irrigation system can solve the problems of using conventional energy resources for running the irrigation system and also prevents the water wastage by continuous monitoring of the field.

Keywords Drip irrigation · Renewable energy · Solar panel · Sensor · Microcontroller

1 Introduction

Irrigation is a broad area which means, supplying of water to the growing plants. Irrigation process is also important to maintain landscaping features, such as grass, trees, shrubs, and flowers. Irrigation is very important process since humans first began cultivating plants. Water requires to cultivate plants in order to grow and flourish. There are so many regions where the rain water is not sufficient to achieve this goal. There are many areas which face drought for short or long period of time. Irrigation is the right way to sort out these problems. Based on factors such as type of crop, soil Topography, quantity and quality of water in soil, climatologic conditions, availability and reliability of energy, irrigation processes are selected. Irrigation methods are of three types.

Surface Irrigation: In this process water is applied to the ground by flowing out a small flood.

Sprinkler Irrigation: Here water is sprayed over the ground as rain.

Drip Irrigation: In this method water is applied to plants in very limited amount which is approximately equal to the rate, that absorption of water by soil and consumption of water by plants.

But surface irrigation and sprinkler irrigation methods have so many disadvantages. Such irrigation processes are not compatible for irregular slopes. For this case considerable land levelling first required. In this method the use of water is in huge amount as water loss is occurred due to evaporation, transpiration and water erosion. Initially, it will be expensive to construct a system and it requires some skill to operate. Wind can push the spray away from the plants in sprinkler method of irrigation. A considerable amount of water can be wastage by evaporation.

2 Automated Drip Irrigation System

Agriculture is the most important part for the growth of any human civilization. There are so many research works have been done to improve agricultural process. Some of the parameters in environment like moisture content of soil, temperature of Environment, humidity, pH, solar radiation, etc. are the major responsible cause for the overall growth of the plant. Many of plant activities such as pollination, germination, etc. can be affected by temperature. In many research work, it is reported that at high temperature, rate of respiration of plants will be high reducing the sugar content of fruits and vegetables. Again at lower temperature the rate of photosynthesis decreases [1]. Humidity of environment can control the moisture loss and temperature management of the plant. If the amount of water molecules are higher in environment the rate of evapo-transmission decreases and the extra water molecules saturate on the surface of leaf resulting expansion and fungus formation on leaf. Another important parameter is moisture for seed germination and intake of nutrients by the plants. Additional water may interfere the interchange of gaseous molecules in between soil and the atmosphere which decreases the rate of respiration as well as growth rate of root. The ideal level of water content in soil is also very important ensuring healthy growth of the root and overall development of the plant [2]. There should be a development for the best solution to maintain the balance in between these parameters and environment. So we need a continuous observation for efficient running of the system. The traditional methods, like sprinkler method and flood type method, are not that much efficient.

The irrigation is a process where we water the soil, important for plant's growth. Drip irrigation is also called by micro-irrigation. This process can reduce the consumption of water as well as fertilizers by dripping the water drop by drop directly at the root of the plants or on the surface of the soil onto the root zone. Thus large quantity of water, used up, will be saved and also use of fertilizers can be reduced as this are carried by water to the plants. In India, agricultural science totally depends on rain water which is not a reliable source of water. So an automatic irrigation method should be applied which will offer water molecules to the plants after checking the condition of moisture level of soil, temperature and soil types & fertilizers. The amount of water containing in soil is called by soil moisture. For a big farm, having horticulture activities, automated irrigation process is the only way to water the plants. It is very important for plant growths. In this automated drip irrigation system, the essential fertilizers can also be applied by this method very effectively without any wastage. The traditional irrigation method causes wastage of a lot of water and also responsible for the fungus formation in plants due to the presence of excessive water molecules. The automated irrigation method can sort out all these problems. The wireless sensors can control all the environment-related parameters for developed irrigation method. Thus this method can conserve the water resource from wastage. Around 85% of total available water in earth is used for agriculture worldwide [3]. With the increase in population, the requirement for water is also increasing in upcoming years. To satisfy the demand we should think about the new techniques for irrigation purpose which can conserve more water. In this automated irrigation process sensor will monitor the soil moisture and water the plants as per requirements. Thus this process can also control the water consumption. The technology for irrigation process has improved in such a way that it became more economic for the farmers. Wireless sensors are used in this system. To enhance the activity of sensors, they are distributed in many areas of the agricultural field which will monitor all the parameters controlling the growth of plants remotely. In wireless sensor network, a sensor is incorporated and consists of small nodes which work on its own. The aim of using this nodes is to collect all the data from the field and transmit it in a central system where all the data are collected. Various types of sensor protocols have been used in many research works to form a network of wireless sensors.

2.1 Literature Review

The development of technology is important in this field to improve the crop generation and decrease the manpower requirements, consumption of water as well as fertilizers. So many research works are going on for the betterment of farmers providing developed techniques to increase the agricultural yield.

In Turkey, approximately 75% of the pure water is spent for the irrigation method in agricultural [4]. Hence management of water resources effectively is essential and important in this case [5, 6]. Farmers in dry zones are presently using new irrigation practices to get more yield per drop by using less freshwater [7, 8]. Here different types of sensors are used [9]. This automated sensor-based irrigation system is an effective way to increase crop production and efficacy of grounds to use less amount of water during irrigation process. Due to the different soil characteristics, the amount of available water will be different and thus site-specific controlling irrigation valves are required [7, 10, 11]. The sensors can make decision that when or where to irrigate and how much water should be used for irrigation. The moisture level of soil during the different stages of plants growth should be observed to reduce water stress and also it can improve crop yields [12, 13]. It has been reported that there are so many cases where sensors are used in agriculture [10, 14–20].

In the last two decades, there are so many research works have done on automated irrigation process using sensors in agricultural systems [21, 22]. In a research work, for automated irrigation method micro-sprinkler system is used for latching, where a valves are used in citrus orchard field with sensors [23]. Moisture content sensors of soil and sprinkler controllers are being introduced for this type of process [5, 24–27]. The use of wireless network can reduce cost of the electric wiring and very easy to install and maintain in large area [28–30]. After so many uses of wireless technologies in different cases, a trial was made involving different equipment's in the perspective of controller [11, 24, 31]. In a research work, microcontrollers were also used in this type of irrigation method and this wireless automated system was applied with a field programmable gate array (FPGA) [32].

There are so many entities should be implemented in precision agriculture for different utilities [29, 33–35]. Again there are so many research works which have been done so far where microwave remote sensing is used in this irrigation process [36–38]. But the assembling of sensors, valves and modules are too complex to be feasible and due to cost, installation time, maintenance, and complexity of systems, there are few cases where the process has been used [39, 40].

2.2 *Methodology*

2.2.1 Moisture Measurement

Soil contains so many different elements like mineral and organic particles. Water and air molecules are also there. Soil can be mainly classified into following 4 classes, which are Clay (0–0.002 mm), silt (0.002–0.075 mm), Sand (0.075–4.75 mm), Gravel (4.75–80 mm). All categories have different properties and thus water holding capacities will be different.

While water molecules penetrate into soil particles (Fig. 1a), it removes the air molecules and fills up the void spaces remaining in soil. When all void spaces are filled with water, the stage is called by saturation point. But this stage remains stable for a very short period of time. Due to gravitational field, the water molecules percolate downward following the water profile. Simultaneously by the capillary action the downward movement of water molecules is hindered and a balanced condition is reached. Thus, field capacity is defined as the step, when void space is filled with water and air molecules (Fig. 1b). Every crop has a Critical Soil Moisture Deficit level. Beyond this level, if the soil is dried out, the production of crops will decrease. Afterward extra water molecules are removed from soil particles and less amount of water is hold by the soil particles due to the surface tension. This situation is called by wilting point (Fig. 1c). The amount of available water can be determined from the difference between amount of water at field capacity and the amount of water stored at the permanent wilting point. Most of the available water required for the plants is coming from the upper portion of root zone.

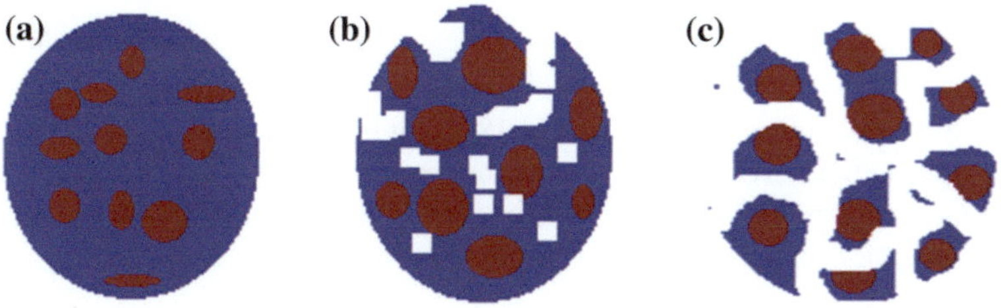

Fig. 1 Water holding capacity of soil. **a** Excess water, percolation of water; **b** Water available due to capillary action; **c** Unavailable water

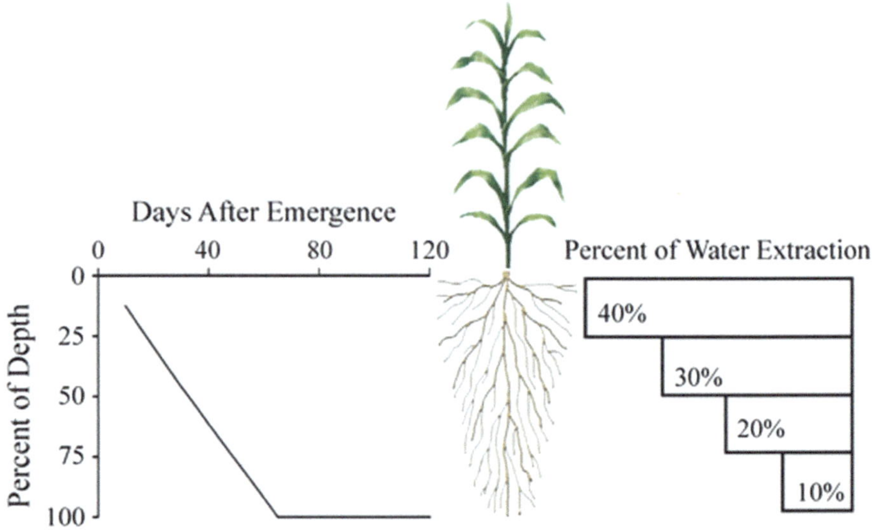

Fig. 2 Effective root zone Depth

The effective root zone depth is referred as the depth of the soil, where roughly 70% of water is stored for the plants (Fig. 2).

There are so many methods to measure the level of soil moisture.

Gravimetric Method

In the year 1894, gravimetric method is being introduced to determine the soil moisture [41]. In this technique, first step was to collect soil sample. Then the weight of the sample was taken followed by drying in oven for 24 h. After that the sample was again weighted. The ratio of two weights are given the % of moisture content. But this method is very time-consuming and labor cost involved for collecting the sample.

Electrical Conductivity Measurement

Though this process was first reported in 1897 but in 1940, a successful demonstration was given after development of electrical unit [41]. This electrical conductivity measurement method followed the principle of conductivity. When the resistance is high, the presence of water molecules in soil is less.

Tensiometer

Tensiometer measured the soil moisture tension for the first time in 1936. In this process a tube is filled with water molecules, the bottom end of which is porous. Thus the direction of water flow is in both side. At the other end, a pressure meter is connected. At low moisture level, water flows to the soil from the tube creating vacuum which is indicated by an increase in pressure and vice versa. This tension change is measured [42].

Radioactivity

In 1950, radioactive method was first introduced, using neutron probe [42]. The emitted neutrons are slowed down by the atoms having low atomic weight. When the amount of water is high in soil, a large number of hydrogen atoms are present which are light weight. Thus it can slowed down the neutrons. This neutrons are calculated and converted into electricity. For this process a skilled personnel is required as neutrons are radioactive. Hence this method is slow and costly.

Time Domain Reflectometry

This process is based on the transmission of electromagnetic (EM) waves. The propagation of electromagnetic wave (EM) through soil particles affected by the moisture content and electrical conductivity of soil particles. Then the passing EM through the transmission lines is collected. The change in propagation constant (i.e., velocity and attenuation) indicates that how much water is remaining in soil particles.

Plant Canopy Temperature Measurement

In this method, the thermal imaging of the plant canopy was taken. Initially the image of crop is taken and collected. After that from the same position the image of the crop is taken by IR imaging process and then these images are super imposed to measure the water shortage area of the canopy. From all these data, water pressure index is calculated and automatic irrigation is being proceed as per requirement. This method is very effective and important to determine the moisture content in soil.

2.2.2 Temperature Measurement

Temperature is an essential parameter for the growth of plants in agriculture. Hence continuous monitoring of temperature is required for an effective irrigation process. There are so many methods which have been applied for the measurement of many physical properties dependent on temperature. Among these techniques some of them are thermocouple, thermistor, RTD, pyrometer, Langmuir probes, infrared, etc.

2.2.3 Humidity Measurement

Humidity is defined as the amount of water vapor in air. Again absolute humidity can be defined as the quantity of moisture in a specified volume of air and relative humidity is defined as the ratio of moisture in the air to the maximum amount of moisture that the air can hold. It differs with the change in air temperature like hotter air, can hold more moisture.

2.2.4 Wireless Sensor Network

There are different techniques used to measure soil moisture content. Broadly those can be categorized as, a. modern techniques b. classical techniques.

a. To measure the amount of soil moisture we generally use soil sensor, tensiometers, moisture balance equipment. Time Domain Reflectometry (TDR), Frequency Domain Reflectometry (FDR) capacitance technique, and optical techniques are well known as modern technique [43].
b. Some examples of the classical techniques in this field are thermo-gravimetric method, calcium carbide neutron scattering method, gypsum block and tensiometer methods.

By using sensor water content is checked in real time. Soil moisture sensor also makes it possible to water the field based on crop and field characteristics. Above-mentioned sensors can work as "stand-alone" or it can combine with FAO method [44].

WSN has gained much attention in recent days due to its distributed monitoring capability. This technique needs low cost, less power and this self-sustained systems spread across entire field for monitoring and reporting data in real time. There are three main units for WSN-1. Monitoring sensor unit (one or more) 2. Processing unit 3. Wireless transceiver. Processing unit will have main power resource as solar, vibration, fuel cells, acoustic noise and mobile supplier for replenishment and back up source such as battery. Solar energy is the most popular source among primary sources [45].

There are two types of sensor nodes-

1. Generic node—This type of node embeds multiple sensors.
2. Gateway node—This collects data from generic node and sends it to main unit.

There are five categories of WSN based on its deployment. They are terrestrial, underground, underwater, Multimedia and Mobile WSN [46]. There are many sectors where WSN is used such as military tracking and surveillance, biomedical health monitoring, inventory tracking, hazardous environment exploration and seismic sensing. There are few issues associated with WSN, first one is related to its infrastructure which can only work with selective platform, operating system and storage systems to create different user related applications. Communication protocol is wireless and not conventional. Conventional protocols requires more resources and power and are not suitable for wireless node. If low cost, low resource and low power is required, then new protocol needs to be developed. Some of the examples are Bluetooth, ZigBee [IEEE 802.15.4], WirelessHART, etc. [8].

Field Application of Wireless Sensor Network (WSN)

Nowadays a new single probe heat pulse sensor (SPHP) was developed by Dias et al. [47]. In this SPHP, a transistor is there which can work as heating and temperature sensing element.

In china, Xiao et al. [48] designed a sensor for paddy field, which is called by wireless, integrated, frequency domain soil moisture sensor (WFDSS). This was used to monitor moisture content of soil and water depth simultaneously and send data without any wire to remote data management center.

Montoya et al. [49] developed an Android application that helps the farmer to ON/OFF the motor remotely in the field.

Bartlett et al. [50] experimented in a slope mountain catchment area in four places in southern Italy, with multi-frequency sensor GEM-300 for the monitoring the soil moisture content. Six different frequencies were tested to get the optimized frequency at which GEM-300 sensor could be used for the recovery of soil from being dried throughout the year.

Cardenas-Lailhacar and Duke [51] tested with common bermuda grass using sensors, that was kept underground at 7–10 cm depth on plots. A calibrated ECH_2O probe was used to monitor continuously water content of the soil.

Blonquist et al. [43] used sprinkler system with the Acclima Digital Time Domain Transmission Sensor for irrigation control and scheduling. This system used 16% less water compared to evapotranspiration estimated previously from a weather station. Also by keeping a fixed irrigation rate of 50 mm per week (7.14 mm/day), 53% less water was applied.

Vellidis et al. [34] used the set of watermark wireless smart sensor to measure soil moisture and soil temperature. A large quantity of sensors were installed in a cotton field and collected data from that sensors was transmitted wirelessly to a centrally located receiver from the sensors.

Shabadi et al. [52] designed an online evapotranspiration scheduling tool for irrigation. This system works with the help of two weather stations.

Micro Drip Irrigation System was applied in a field where citrus (orange) plant was there. Size of the field was over $18 \times 66\ m^2$ and contains 44 plants. Real-time environment parameters were collected from sensor node. The parameters that are monitored were soil moisture, temperature, and humidity. Central node was designed to set the action based on data collected on settings feed earlier. Setting can be changed

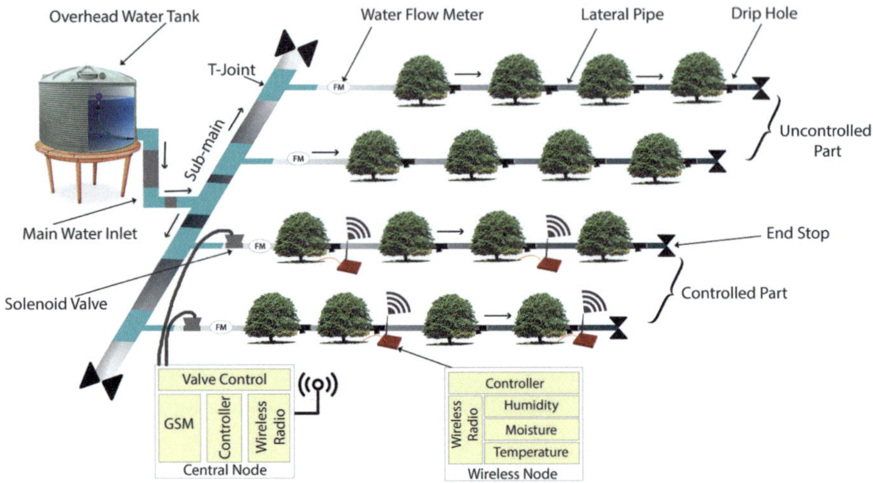

Fig. 3 Automated irrigation system

later any time. The data is stored in memory card with time stamp to log data and users will be notified via GSM state of the firm. One android app will be required user end to access the data and monitor farm condition.

Below diagram Fig. 3, design of WSN system has been shown. A hierarchical architecture been shown by Sathya et al. Sensors were used to monitor temperature, humidity, and luminosity. These sensors interface with digital sensors and take data as per the requirement. As the range of radio frequency of wireless communication is above 1 km, the sensor nodes can be widely separated. Wireless link is used to transmit data from sensor 1 & 2 to sensor 3. The coordinator aggregates data from sensor 1 & 2 as time bound way. Network formed by three sensors called patch network. Sensor node coordinator helps to monitor the total field area by the WSN and directly linked with the gateway node. Data is then transmitted to the remote computer terminal from sensors. The computer terminal has capability to log data related to the measurement of temperature, humidity, and luminosity.

The automatic plant irrigation system can improve agriculture productivity and also capability to monitor agricultural practices. The above-mentioned techniques can be implemented using GSN Bluetooth to control the irrigation process Fig. 4. The sensors installed for the monitoring of crops growth, are controlled through SMS using a GSM module. SMS shared by BLUETOOTH or GSM technique interfaced with the main microcontroller chip. Microcontroller controlled the required process in the field.

This system can also monitor the temp. It includes a 8-bit microcontroller chip, a GSM, and Bluetooth module as well as microcontroller which is used for different purpose to monitor the crop growth. One analog-digital converter is also used to convert the generated analog data to digital. Then SMS is sent to subscriber mobile through GSM and Bluetooth [53].

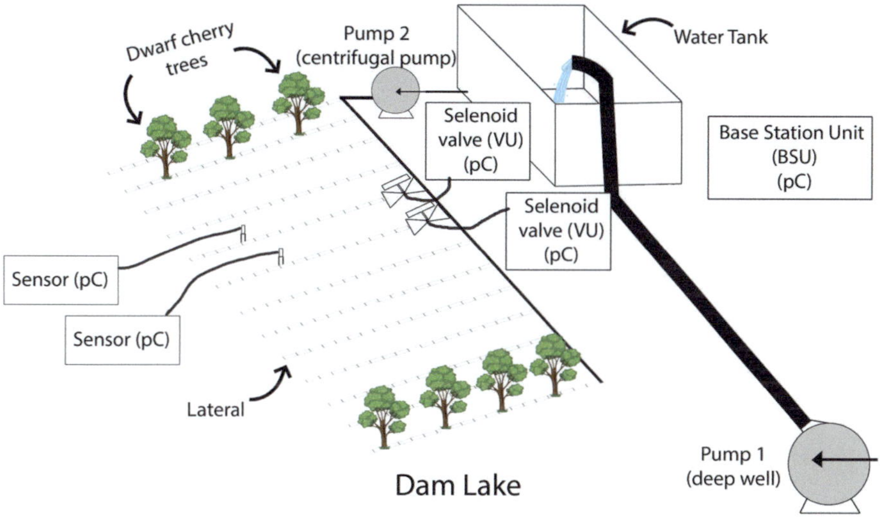

Fig. 4 Overview of system

Irrigation is one of the important parts of agricultural activity and very important for crop growth in the world. In countries like Turkey, approximately 75% of freshwater is used for agriculture. Efficient water management system is very critical for agricultural economy of a country. In the regions, where deficiency of water is there, the farmers should apply some developed techniques for irrigation process to produce more crops [54].

Using wireless sensor-based irrigation control system can optimize yield and improve usage of water for thee fields. Based on the variations of the soil character and crop type, water supply can be managed using control irrigation valve.

An optimum water control system was designed, using motor pump. This system can provide required quantity of water at specific time intervals as per climate condition. There are many features of this system, which vary with parameters and alert the farmers through keyboard/SMS/number of miscalls.

The scheme is showed, Fig. 1, in a block diagram. Based on the commands received, the system sends signals to switch on the motor. Three phase induction motors are chosen for agriculture purpose and they are working on Direct-on-line Starter are.

Prime Objectives of New Generation Agriculture System

- **Remote monitoring**:

    (a) Data collected by gauges and sensors (soil moisture, pressure, environmental, etc.

(b) Status of irrigation valves
(c) Condition of pumping equipment.

- **Remote Control**:
- **Information transfer**:

(a) Automatic addition of environmental data into decision support systems and crop models
(b) Uploading maps to variable rate application equipment
(c) Weather, market & operational information
(d) To remote locations & vehicles
(e) Real-time information such as DGPS correction signals.

- **Communication**:

(a) These systems are typically designed and installed for different purposes which can send text, graphical, voice, and video messages between peoples.
(b) Using embedded system drip irrigation:
(c) Soil moisture content sensors, temperature monitoring sensors, signal conditioning circuit, digital-to-analog converter, LCD module, relay driver, solenoid control valves, etc. are assimilated in the automated control system. The unit is shown in Figure. The indication sent by the sensor is amplified to the required level through corresponding amplifiers used in this system. Then the amplified signal is passed through A/D converters of desired resolution to obtain digital form of input signals [55–58].

A procedure is here implemented for ideal condition of water supply in the agricultural fields. The growth of plants will be proper by distributing required quantity of water at a specific time gap depending on the climatic condition. This system incorporates some of the developed features of automatic control, which is based on the parameters. These parameters are already specified. This system can alert users in case of abnormal conditions like power failure, dry-running, etc. The completion of job can be indicated by buzzer/miscall. The features of the system can set by keyboard and the condition of environmental parameters also can be received through SMS/number of miscalls in specified duration from user mobile to control system cell phone through serial cable (Model Nokia 6610). Whatever signal received from previous step, microcontroller system takes the decision to run the motor through starter which is controlled by its ports. The induction motor is working at 5, 7, 10.5 hp and is chosen for agriculture purpose.

While the drip irrigation system and android system are connected through Internet, the application starts its execution. For setting up this connection, an authentication procedure takes place between the two. Here microcontroller stores the data of soil condition collected from sensor. These data are shown in the software representing as 3D plots. The 3D graphs generated from the system are helpful to indicate and visualize the entire field condition and thus a decisive action can be taken as per requirement.

Benefits of Sensor-Based Irrigation

The primary goal of this system is to apply an optimum amount of irrigation throughout the fields. The irrigation based on sensor has the potential to control the use of water and also economic efficiency by reducing the cost of inputs and increase the yield for the same output. In the traditional method, the huge number of human resources is required. But in this automated sensor-based irrigation method, the requirement is minimized [59]. The benefits of this technology can be seen as follows:

- Water Saving: By this method the wastage of water can be controlled in many research work [60]. It has been reported in so many study that the use of sensors with drip and sprinkler irrigation system can reduce the use of water up to 80–90% as against 40–45% in surface irrigation method [61]. Blonquist et al. [62] concluded that 16% of water could be saved in this irrigation practice.
- Yield and Profit: It is very important factor for better production of crops that irrigation process should run in proper time gap. Little delay in irrigation can result a loss in between US$62/ha and US$300/ha [63]. Thus it is shown that agricultural yields were increased and a loss of 2 billion pounds were recovered in a year. This technology can be used to improve water use efficiency in semiarid region like Rajasthan. In Rajasthan the groundwater is consumed for all sort of domestic and industrial work. But the level of groundwater is decreasing by 1 m per year. Hence the old irrigation system leads over utilization of water in an area and in another area; it causes the lack of irrigation. This results lowering of water use efficiency and water application is not optimized. Farmers irrigate their fields based on their traditional knowledge without taking the actual amount of soil moisture level in the field, soil type, crop water requirement and weather forecasting. If farmers of this region will use this sensor-based technology, then they could be advised about irrigation scheduling in their field and could increase their water use efficiency and yield. Thus it can solve the problem of food of this state as well as our country.

3 Solar-Powered Automated Drip Irrigation System

The main economy of India is based on agriculture and also it is important for survival. Recently two major problems have been reported in Indian agriculture. Nowadays government has taken an initiative to supply free electricity for farmers to run their water pumps. But constantly the farmers are misusing this facility by running their home appliances like TV, motor, fans, etc. Thus the free supply of electricity become a major problem for the government. Again most of the farmers have very little knowledge about their field, condition of soil, and its fertility of their field. They even cannot find the right seeds for their fields to be sown.

Around the world same condition is shown. Kazakhstan is a large landlocked nation that has a largely agrarian economy. One quarter of its residents makes their living by farming. Many residents in the town of Shelek, Kazakhstan depend on small market gardens to provide their families with food and additional income that can be obtained from selling their surplus crops at local farmer's markets, but in Shelek there is no infrastructure to make electrical grid. In addition, those who can afford to have electricity lines to run their houses experience inconsistent power due to high wind speeds that plague the area. When winds reach upward of 70 mph the electrical grid is shut down. Thus all the families in this area face a huge problem in irrigation process. Due to inconsistent electricity, some well-water extracting pumps are not working any more.

The most easily available energy resource in world is solar energy. The use of solar energy is not only economic way, but also it is environment friendly. Nowadays solar energy is used in many cases like street light, domestic purposes due to its renewability. Thus in automated irrigation process solar energy is also used. In India conventional energy sources are the major problem. Hence the best solution of Indian farmers is the use of solar energy. Due to the over consumption of water for irrigation process in an unplanned way, the groundwater level is decreasing causing many regions as an irrigate area. Nowadays due to huge growth of population, the demand of food also increases. Thus the balance between demand and supply of food is destroyed. To make a balance in between demand and supply of food, the yield of crop should be increased. In automated irrigation system, use of solar power offers a very simple and economical solution to this problem.

3.1 Related Work

There are three parts of SCADA irrigation system SCADA, PLC, and solar panel in Fig. 5. SCADA known as the supervisory control and data collecting unit gathers and analyzes time data. SCADA can control and monitor the expansively distributed system from an extended distance. The automation part of this system is managed by Remote Terminal Unit (RTU) and Programmable Logic Controller (PLC). SCADA system has hardware and the software parts. Information from input device is collected by Remote Terminal Unit (RTU).

Information is kept in memory of master terminal unit (MTU) and then transfer that to master terminal unit (MTU) [34]. MTU collects and sends the data to the device. MTU can be a computer for processing data and information [64]. Communication connection is critical in SCADA system. It sends signal for processing the 56,786 data and enables system to make decision for working according to the data. LAN and wireless radio signal are used [65]. NI LABVIEW software is used to communicate with a Programmable Logic Controller (PLC). Modbus serial communication protocol was used to communicate successfully with PLC, made by Modicon in 1979, and it was then expanded to the TCP protocol. Now Modbus is used in the

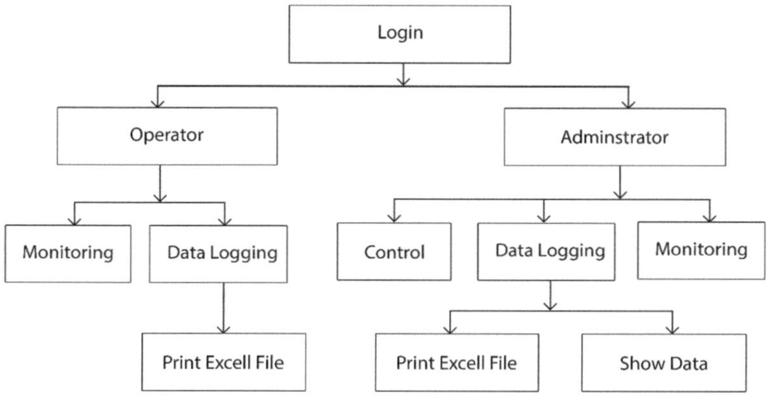

Fig. 5 SCADA system over view

industry widely as communications medium. Modbus can assist in networking. It turns security into the user interfaces [66].

A agricultural system, run by solar energy, Fig. 6, is sustainable and economical for crop production, as it is run by solar energy [67]. System is designed for peak water demand between 7.30 a.m. and 3 p.m. The time period between sunrise and sunset is called peak production time [68].

System is designed to work based on time and has capability to predict weather condition. If the farmland located in an area which has irregular patterns of rain, then the field might get over watered. Water releases to the seedlings in trays, which can result in dislodged seedlings from trays. Thus this process can lower the rate of crop generation and increase the labor cost [68].

Due to some issues with solar system, the alternative systems as mechanical windmill have been experimented [69]. Solar energy, energy generated from mechanical windmills or portable generators, can be used where electricity is not available. This system reduces load in stream banks and reducing the factors like nutrient filling, streamside erosion, vegetation, and pollution of environment [70]. Though initially the installation cost of a solar system is high but the maintenance charge is low, compared to other system. Fixed mounts can resist in high wind and it is also less expensive. A big tank is used to store the enough quantity of water. Thus it can reduce the maintenance as well as installation cost [70]. There is a controller that helps the pumps to run at low as well as high voltage [70].

In the above solar-powered irrigation system Fig. 7, the farmers grows vegetables on the slopes which is made artificially and thus these slopes are remaining green during the dry seasons also. Here photovoltaic cell is used for producing electricity. This electricity is stored later in batteries. After that the power is provided by these batteries to run the system successfully. There is a big tank to store the water and from this storage, water is taken up to the sprinklers by an irrigation pump. The sprinklers are fitted at the slope toe for watering the plants.

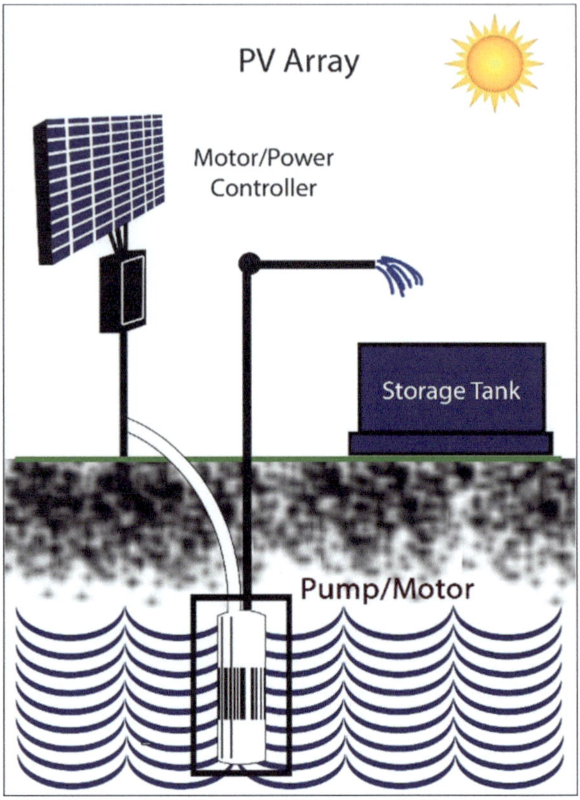

Fig. 6 Photovoltaic system

India is a country, where so many research works have been done to develop the conventional agricultural practices for the better productivity. In another research work, the solar panels can drive the water pump from the resource [70]. Though some of the drawbacks are there to use solar energy, like photovoltaicity for water pump, still it is environment friendly [71]. Solar panels (an array of photovoltaic cells) are nowadays widely used for street lights, water heaters, and other domestic works. Compared to the previous era, the cost of solar panels is reducing gradually. It is a perfect alternative of all energies as in India energy crisis is a serious problem [72]. There are so many problems like low rate of crop production and less availability of water for agriculture can be solved by using this proposed solar-powered automated irrigation system.

Solar panels, Fig. 8, are used in the whole system. In this case solar panels of 20 V, 10 W is used which have output current of 0.5 A. With the help of solar energy, the electrons are getting excited to conduction band and thus produce current. This DC current is stored in the battery (12 V, 1 A) and 7815 voltage regulator is also used. Thus the pump can be operated even in the night time. The battery is connected to the ARDUINO UNO with other end.

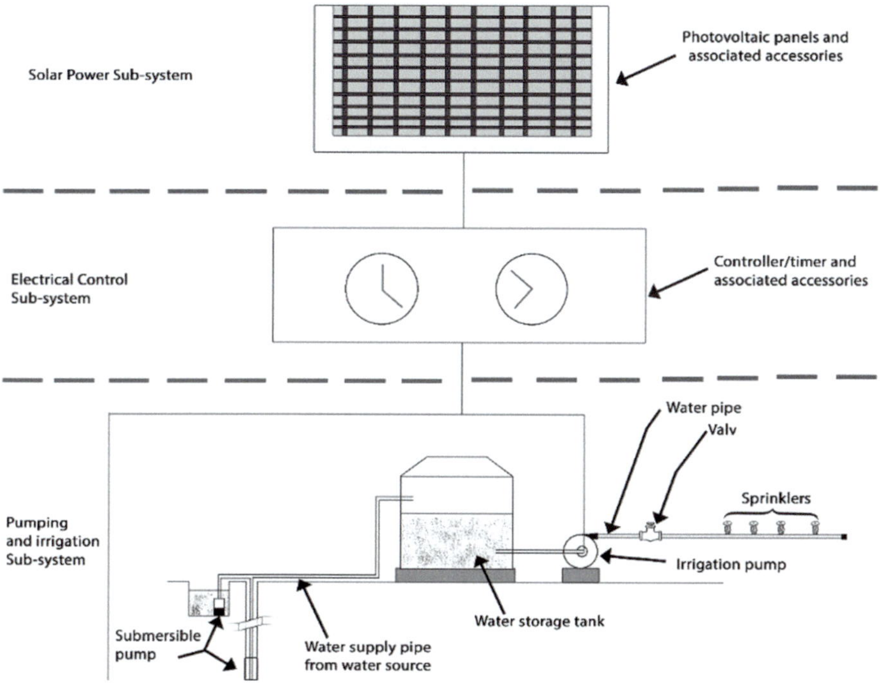

Fig. 7 Solar Installation System

Fig. 8 Solar panels

In this system, a sensor, a water pump, and an ARDUINO UNO are used. ARDUINO UNO is a microcontroller board based on ATmega328, where ARDUINO IDE software is used. ARDUINO can control both motor and moisture sensor. The moisture sensor determines the water content in the soil and sends the data to the ARDUINO. The threshold limit of the moisture level is predefined. If the level of moisture is below the threshold limit, it indicates that the moisture of the soil is less than ideal value. Then ARDUINO controls the motor to run and finally water will reach to the plants. The water level determining sensor is also used to monitor the water level in storage tank. If the water level of the tank is not sufficient to water the plant, then sensor will help the motor to run for more time. If water level is less than the certain threshold value, then the motor will start else motor is OFF. The value of moisture level can be displayed on WI-FI module ESP8266.

In the photovoltaic system, solar cells are used which absorb the sunlight and convert the energy into electricity. The photovoltaic cell is made of silicon or any other semiconductors which converts the solar energy into DC electricity through photovoltaic effects. When the sunlight interacts with the solar cell, it absorbs solar energy and generates electricity later. When the sunlight interacts with the semiconductor, electricity will be produced.

ARDUINO UNO

It is an open-source microcontroller compatible with this system. This microcontrollers are not very expensive. They even use low electrical power around 5.5 V. C and C++ were the working language for this system. ARDUINO is connected to a computer via the universal serial bus (USB). The ARDUINO, Fig. 9, is a microcontroller platform, mounted on a board that can be incorporated very easily in most computers. With the help of ARDUINO, the user can do various things by using different programming language, which is called sketches.

Fig. 9 Arduino UNO

Fig. 10 Soil moisture sensor

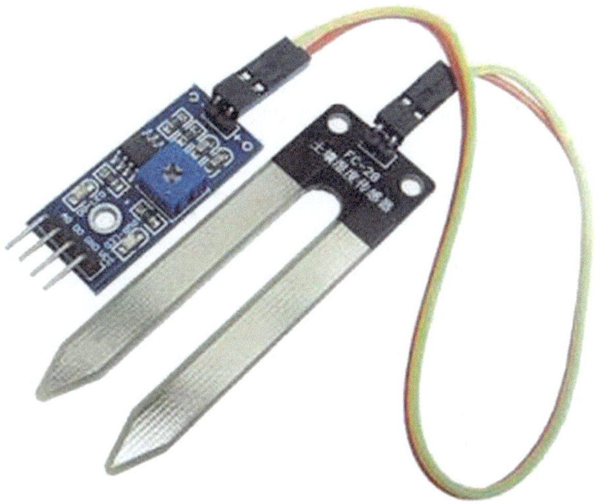

Soil moisture determining sensor, Fig. 10, is used to monitor the amount of moisture in the soil. The required amount of moisture of soil for better crop generation is defined in the microcontroller previously. The moisture sensor follows the principle of resistance. Water is a good conductor. When the moisture level of soil is low, the resistance is high between the electrodes. But if the moisture level is high, then water molecules help to conduct electricity. When the moisture level is lower than the pre-set value, the pumps get on. After that when the water level reaches the required value, the pump is cutoff as the power supply is stopped.

Batteries are connected in series. It stores the solar energy and supply the electricity, generated by solar cell. The power of the battery used in this system is of 12 V, 1 A. 18 h is required to charge the battery completely by the solar panel used for this system.

4 Conclusion

The greatest advantages of using solar energy is its availability and it is also a renewable source of energy. For this type of automated irrigation method, a huge knowledge is required about renewable non-conventional energy and also about the environmental parameters which control irrigation system. Hence the system is conserving the energy resources and human resources. 10 Through this irrigation process, the monitoring of environmental condition of field can be done remotely. Nowadays modern technology is incorporated in this irrigation to low the cost. In automated drip irrigation system, power consumption is very less and it completely relies on renewable solar energy.

References

1. Plant Growth Factors: Temperature, Colorado State University, Available (as on 14-09-2015) at: http://www.ext.colostate.edu/mg/Gardenno-tes/143.html#heat
2. Plant Growth Factors: Water, Colorado State University, Available (as on 14-09-2015) at: http://www.ext.colostate.edu/-mg/Gardennotes/144.html
3. Bigas, H. (ed.): The global water crisis: addressing an urgent security issue. Paper for inter action council, Hamilton, Canada: UNO-INWEH, 2011–12
4. Dursun, M., Ozden, S.: A prototype of PC based remote control of irrigation. In: International Conference on Environmental Engineering and Application (ICEEA), pp. 255–258. (IEEE Catalog Number: CFP1020L-PRT) (2010)
5. Kim, Y., Evans, R.G.: Software design for wireless sensor-based site-specific irrigation. Comput. Electron. Agric. **66**, 159–165 (2009)
6. Sezen, S.M., Celikel, G., Yazar, A., Tekin, S., Kapur, B.: Effect of irrigation management on yield and quality of tomatoes grown in different soilless media in a glasshouse. Sci. Res. Essay **5**(1), 41–48 (2010)
7. Balendonck, J., Hemming, J., Van, T. B., Pardossi, A., Incrocci, L., Marzialetti, P.: Sensors and wireless sensor networks for irrigation management under deficit conditions (FLOW-AID). In: International Conference on Agricultural Engineering (AgEng2008). Conference Proceeding, p. 19 (2008)
8. Ngaira, J.K.W.: Impact of climate change on agriculture in Africa by 2030. Sci. Res. Essays **2**(7), 238–243 (2007)
9. Lopez, R.J.A., Sotoa, F., Suardiaza, F., Sancheza, P., Iborraa, A., Verab, J.A.: Wireless sensor networks for precision horticulture in southern spain. Comput. Electron. Agric. **68**, 25–35 (2009)
10. Kim, Y., Evans, R.G., Iversen, W.M.: Remote sensing and control of an irrigation system using a distributed wireless sensor network. IEEE Trans. Instrum. Meas. **57**(7), 1379–1387 (2008)
11. Coates, R.W., Delwiche, M.J., Brown, P.H.: Control of individual micro sprinklers and fault detection strategies. Precision Agric. **7**, 85–99 (2006)
12. Doraiswamy, P.C., Hatfield, J.L., Jackson, T.J., Akhmedov, B., Prueger, J., Stern, A.: Crop condition and yield simulations using Landsat and MODIS. Remote. Sens. Environ. **92**, 548–559 (2004)
13. Coates, R.W., Delwiche, M., Brown, P.: Precision Irrigation in orchards: development of a spatially variable micro sprinkler system. In: Information and Technology for Sustainable Fruit and Vegetable Production (FRUTIC), pp. 611–624 (2005)
14. Jacobson, B.K., Jones, P.H., Jones, J.W., Paramore, J.A.: Real-time greenhouse monitoring and control with an expert system. Comput. Electron. Agric. **3**, 273–285 (1989)
15. Meron, M., Assaf, R., Bravdo, B., Wallach, R., Hallel, R., Levin, A., Dahan, I.: Soil sensor actuated micro irrigation of apples. In: Proceedings of the 5th International Micro Irrigation Congress, ASABE, pp. 486–491 (1995)
16. Stone, K.C., Smajstrla, A.G., Zazueta, F.S.: Microcomputer-based data acquisition system for continuous soil water potential measurements. Soil Crop Sci. Soc. Fla. Proc. **44**, 49–53 (1985)
17. Testezlaf, R., Zazueta, F.S., Yeager, T.H.: A real-time irrigation control system for greenhouses. Appl. Eng. Agric. **13**(3), 329–332 (1997)
18. Zazueta, F.S., Smajstrla, A.G.: Microcomputer-based control of irrigation systems. Appl. Eng. Agric. **8**(5), 593–596 (1992)
19. Abreu, V.M., Pereira, L.S.: Sprinkler irrigation systems design using ISAMim. pp. 022254 (2002)
20. Kim, Y., Evans, R.G., Iversen, W.M.: Evaluation of closed-loop site-specific irrigation with wireless sensor network. J. Irrig. Drain. Eng. **135**(1), 25–31 (2009)

21. Oksanen, T., Ohman, M., Miettinen, M., Visala, A.: Open configurable control system for precision farming. Automation Technology for Off-Road Equipment, Proceedings, 701P1004 (2004)
22. Zhang, Z.: Investigation of wireless sensor networks for precision agriculture. In: ASAE Annual International Meeting, pp. 041154 (2004)
23. Torre-Neto, A., Schueller, J.K., Haman, D.Z.: Networked sensing and valve actuation for spatially-variable microsprinkler irrigation. In: ASAE Annual International Meeting, pp. 001158 (2000)
24. Miranda, F.R., Yoder, R., Wilkerson, J.B.: A site-specific irrigation control system. In: ASAE Annual International Meeting, pp. 031129 (2003)
25. Wall, R.W., King, B.A.: Incorporating plug and play technology into measurement and control systems for irrigation management. In: ASAE Annual International Meeting, pp. 042189 (2004)
26. Perry, C.D., Dukes, M.D., Harrison, K.A.: Effects of variable-rate sprinkler cycling on irrigation uniformity. In: ASAE Annual International Meeting, pp. 041117 (2004)
27. Damas, M., Prados, A.M., Gomez, F., Olivares, G.: HidroBus® system: fieldbus for integrated management of extensive areas of irrigated land. Microprocessors Microsyst. **25**, 177–184 (2001)
28. Panchard, J., Rao, S., Prabhakar, T.V., Jamadagni, H.S., Hubaux, J.P.: COMMON-sense net: improved water management for resource-poor farmers via sensor networks. ICTD 2006 Conference (2006)
29. Wang, N., Zhang, N., Wang, M.: Wireless sensors in agriculture and food industry—Recent development and future perspective. Comp. Electron. Agric. **50**, 1–14 (2006)
30. Baggio, A.: Wireless sensor networks in precision agriculture. In: On-line Proceeding of the Workshop on Real-World Wireless Sensor Networks, pp. 50–51 (2005)
31. Coates, R.W., Delwiche, M.J., Brown, P.H.: Design of a system for individual micro sprinkler control. Trans. ASABE **49**(6), 1963–1970 (2006)
32. Mendoza-Jasso, J., Vargas, G.O., Miranda, R.C., Ramos, E.V., Garrido, A.Z., Ruiz, G.H.: FPGA-based real-time remote monitoring system. Comput. Electron. Agric. **49**, 272–285 (2005)
33. Camilli, A., Cugnasca, C.E., Saraiva, A.M., Hirakawa, A.R., Corrêa, L.P.: From wireless sensor to field mapping: anatomy of an application for precision agriculture. Comput. Electron. Agric. **58**, 25–36 (2007)
34. Vellidis, G., Tucker, M., Perry, C., Kvien, C., Bednarz, C.: A real-time wireless smart sensor array for scheduling irrigation. Comput. Electron. Agric. **61**, 44–50 (2008)
35. Siuli Roy, A.D., Bandyopadhyays, S.: Agro-sense: Precision agriculture using sensor-based wireless mesh networks. Innovations in NGN: Future Network and Services. In: Proceedings of the First ITU-T Kaleidoscope Academic Conference (K-INGN 2008), pp. 383–388 (2008)
36. Engman, E.T., Chauhan, N.: Status of microwave soil moisture measurements with remote sensing. Remote. Sens. Environ. **51**, 189–198 (1995)
37. Ulaby, F.T., Dubois, P.C., Zyl, J.J.V.: Radar mapping of surface soil moisture. J. Hydrol. **184**, 57–84 (1996)
38. Jackson, T.J., Vine, D.L., Hsu, A.Y., Oldak, A., Starks, P., Swift, C., Isham, J., Haken, M.: Soil moisture mapping at regional scales using microwave radiometry: the Southern Great Plains Hydrology Experiment. IEEE Trans. Geosci. Remote Sensing **37**, 2136–2151 (1999)
39. Thelen, J., Goense, D., Langendoen, K.: Radio wave propagation in potato fields. In: First Workshop on Wireless Network Measurements (co-located with WiOpt2005), pp. 1–4 (2005)
40. Miranda, F.R., Yoder, R.E., Wilkerson, J.B., Odhiamboc, L.O.: An autonomous controller for site-specific management of fixed irrigation systems. Comput. Electron. Agric. **48**, 183–197 (2005)
41. Johnson, A.I.: Methods of measuring soil moisture in the field. U. S. Geological Survey (1992). (Third Reprint)

42. Morris, M.: NCAT energy specialist, Soil moisture monitoring: low-cost tools and methods. ATTRA (2006)
43. Blonquist, J.M., Jones, S.B., Robinson, D.A.: Precise irrigation scheduling for turfgrass using a subsurface electromagnetic soil moisture sensor. Agric. Water Manag. **84**, 153–165 (2006)
44. Thompson, R.B., Gallardo, M., Valdez, L.C., Fernández, M.D.: Using plant water status to define threshold values for irrigation management of vegetable crops using soil moisture sensors. Agric. Water Manag. **88**, 147–158 (2007)
45. Seah, W.K.G., Eu, Z.A., Tan, H.P.: Wireless sensor networks powered by ambient energy harvesting (wsn-heap)—Survey and challenges. In: Proceeding 1st International Conference on Wireless VITAE, pp. 1–5, May 2009
46. Yick, J., Mukherjee, B., Ghosal, D.: Wireless sensor network survey. Comput. Netw., **52**(12):2292–2330 (2008)
47. Dias, P.C., Roque, W., Ferreira, E.C., Siqueira Dias, J.A.: A high sensitivity single-probe heat pulse soil moisture sensor based on a single NPN junction transistor. Comput. Electron. Agric. **96**, 139–147 (2013)
48. Xiao, D., Feng, J., Wang, N., Luo, X., Hu, Y.: Integrated soil moisture and water depth sensor for paddy fields. Comput. Electron. Agric. **98**, 214–221 (2013)
49. Montoya, F.G., Gómez, J., Cama, A., Sierra, A.Z., Martínez, F., De La Cruz, J.L., Agugliaro, F.M.: A monitoring system for intensive agriculture based on mesh networks and the android system. Comput. Electron. Agric. **99**, 14–20 (2013)
50. Bartlett, A.C., Andales, A.A., Arabi, M., Bauder, T.A.: A smartphone app to extend use of a cloud-based irrigation scheduling tool. Comput. Electron. Agric. **111**, 127–130 (2015)
51. Cardenas-Lailhacar, B., Dukes, M.D.: Precision of soil moisture sensor irrigation controllers under field conditions. Agric. Water Manag. **97**, 666–672 (2010)
52. Shabadi, L., Patil, N., Nikita., M.., Shruti, J., Smitha, P., Swati. C.: Irrigation control system using android and GSM for efficient use of water and power. IJARCSSE **4**(7): 607–611 (2014). www.ijarcsse.com
53. Rashid, H., sahgal J.L., Anshulgangwar Md.R.: Control of irrigation automatically by using wireless sensor network. IJSCE, **3**(1) (2013). ISSN: 2231-2307
54. Balendonck et al., 2008: Ngaira, (2007)
55. Water Management System for Agricultural Sector Prof. Dnyaneshwar Natha Wavhal*, Prof. Sneha Ghumatkar #, Prof. Manish Giri, Prof. Nikhil Kumar B.S.*Dept. of Computer Engineering Jaihind College of Engineering Kuran, University of Pune, Maharashtra, INDIA
56. Rajpal, A., Jain, S., Khare, Nistha., Shukla, K.A.: Microcontroller based Automatic Irrigation System with Moisture Sensors. In: Proceedings of the International Conference on Science and Engineering, pp. 94–96 (2011)
57. "Automatic Irrigation Management System", (October 29, 2011). wwu.edu/changj7/project/docs/proj%20description.pdf
58. "Irrigating basics", published in American nurseryman, December 1 (2007)
59. Kansara, K., Zaveri, V., Shah, S., Delwadkar, S., Jani, K.: Sensor based automated irrigation system with IOT: a technical review. 6(6):5331–5333 (2015)
60. Shah, N.G., Das, I.: Precision irrigation sensor network based irrigation, problems, perspectives and challenges of agricultural water management, pp. 217–232. IIT Bombay, India (2012)
61. Carpena R.M., Dukes, M.D.: Automatic irrigation based on soil moisture for vegetable crops. AE354 (2005). http://edis.ifas.ufl.edu/ae3
62. Gordon, R.C., Deon R., Edwin, W.: Practical modern SCADA protocols: DNP3, 60870.5 and related systems newnes (Elsevier) (2004)
63. Manag, W.: 84: pp 153–165 (2006)
64. Hantula, R.: How Do, Solar Panels Work. (Science in the Real World) (2009)
65. James, M.: Ebeling, Computer monitoring & control technology (Freshwater Institute) (2010)
66. Halvorsen: OPC and real-time systems in labview (Telemark University College) (2012)

67. Irrigation. https://en.wikipedia.org/wiki/irrigation (Wikipedia)
68. Harris: Solar powered automated greenhouse irrigation system (LumiTech International Limited) (2009)
69. Morris, L.: Solar-powered livestock watering systems (ATTRA) (2008)
70. Chitransh, A., Sagar, A., Kumar, A.: Automated solar powered irrigation system a technical review. Int. Res. J. Eng. Technol. (IRJET), **3**(4) (Apr 2016). E-ISSN: 2395-0056
71. Gangadharan, A., Srinath, N., SwathyKrishna, I., Arjunan, S.C.: Solar powered smart irrigation system. IRACST—Int. J. Comput. Sci. Inf. Technol. Secur. (IJCSITS), **6**(2) (Mar-April 2016). ISSN: 2249-9555
72. Kashiv, A., Bilala, A., Shirazi, N., Dwivedi, A., Joshi, R.: Solar Drip Irrigation Syatem. Int J Emerg Technol Adv Eng **6**(4) (April 2016). ISSN 2250-2459, ISO 9001:2008 Certified Journal

IoT Foundations and Its Application

Srabanti Chakraborty, Prasenjit Das and Souvik Pal

Abstract In this age of computing technology, IoT is one of the tools that invokes a world where technology speaks. In this technical era, in our daily life, when it is impossible to think a single moment without a connection with any devices, then IoT is one of the greatest and powerful ways to bind all the devices together through a global connection for giving us a digitalized daily life. Nowadays, IoT has been flourished with its various applications by establishing a remote connection to monitor the real-world smart devices that can govern our daily uses objects to make our lives simpler, safer and technical. In this chapter, we will depict the significant roll of IoT and shows the reason that how IoT is becoming an integral part of our daily life through its applications in various domain.

Keywords IoT · Smart cities · RFID · End-to-End connectivity · Smart devices · Interconnected smart world · Wireless sensors

1 IoT Groundwork

1.1 Introduction

In this era of technology, IoT is an important global network system that ensures a smart world through a global communications and connections between the network devices to generate a sense for collecting a huge data throughout the world and to share that data through the Internet to make a smart world. It is difficult to define the application areas of IoT as nowadays it is an integral part of each and every area of our daily life. The various applications of IoT are changing our daily lives. IoT

S. Chakraborty (✉) · P. Das
Department of Computer Science and Technology,
Elitte Institute of Engineering and Management, Kolkata, India
e-mail: srabanti2k@gmail.com

S. Pal
Department of Computer Science and Engineering, JIS College of Engineering,
Kalyani, India

P. K. Pattnaik et al. (eds.), *IoT and Analytics for Agriculture*, Studies in Big Data 63,
https://doi.org/10.1007/978-981-13-9177-4_3

introduced the concept of smart devices where the connected devices are intended to establish a connection between digital and real world to serve as an interconnected smart world. This chapter will depict the different areas of our daily life where the applications of IoT are flourishing. This chapter of the book deals with the concept of IoT-Internet of Things, the applications of IoT, technologies of IoT and need of IoT.

IoT is an automated system which allows a universal network of interconnected everyday physical objects which are equipped with uniquely addressable devices, embedded with sensors, software, electronics, actuators to connect and exchange data. IoT is a conceptual framework that enables to connect and embed intelligence in devices for establish a machine-to-machine or machine-to-people or machine-to-objects or people-to-machine or people-to-objects connection. IoT helps to collect data from a devices with broad range. IoT invokes a world to us, where technology speaks. In this world, milliard of interconnected technical devices can communicate, sense, converse and exchange information over any private or public IP networks. At this moment, IoT is one of the most important and inventive fields for the researchers. Gradually, Internet of Things is going to be an integral part of the modern technology that ensures to provide a smart and digital life by establishing a communication within different and heterogeneous objects together with us. Internet of Things is hoping to establish the interconnection between the different connected smart devices that embedded with sensors, actuators, software and finally network connectivity to collect and also exchange data.

Internet of things may be considered as the Internet of:

- People to machine
- People to people
- Machine to machine.

The Internet of Things can be described as a gargantuan network system which is consisting of different interconnected network devices through some intermediate technologies. These network devices connect to the network system and gather information through sensors to provide this. The whole process of IoT technology completes through the three essential components: the device, gateway, and cloud that shown in Fig. 1.

Device: The IoT devices help to transmit data from one thing to another through wireless sensors, actuators, software and machines.

Gateway: An IoT gateway is one of the essential and key components of IoT which acts as a linking bridge between sensors, devices and the cloud. The IoT performs numerous important and complicated tasks like protocol translation, not Internet-connected device connectivity and helps to place then in cloud platform.

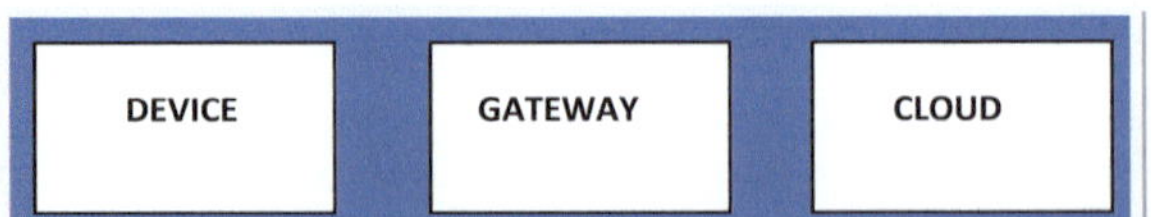

Fig. 1 Components of IoT

Cloud: The cloud platform provides a technique to collect, store and process the information in IoT system.

Figure 2 depicts the total workflow of the IoT with the combination of four phases:

Collection: In this phase, the IoT devices collect the data through sensors from everywhere.

Communication: After collection of data, the communication phase of IoT sends data through a private network system or/and a cloud platform.

Analysis: Analysis phase of IoT creates the information from data and also filtrates the data.

Action: The action phase takes action on the basis of data information and data communication with another IoT device (another machine) and sends report or notification in the form of sms, email or text to talk with another system.

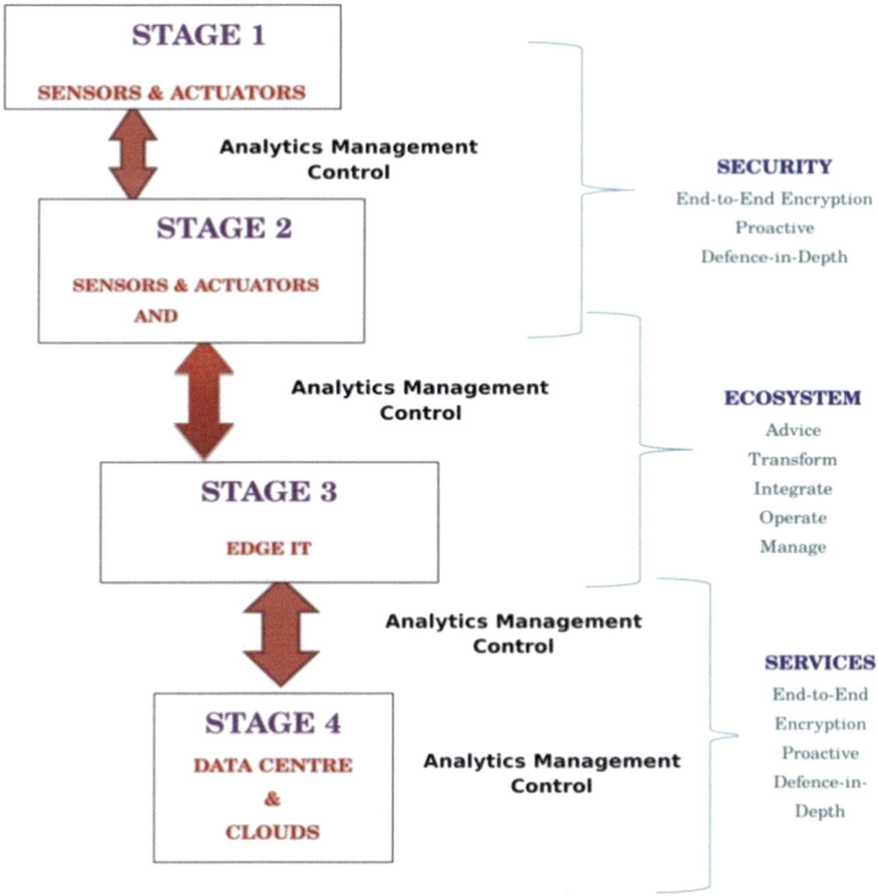

Fig. 2 Workflow of IoT

1.2 Characteristics of Internet of Things

Now IoT is ruling the technical world with some very powerful characteristics, which enables to connect "anything, any service" at "anywhere, anytime". The important and fundamental features of IoT can be illustrated in Fig. 2:

Interconnectivity
This characteristic of Internet of Things ensures that anything that surround us can be physically or virtually connected through a worldwide communication and information system. This feature endows Internet of Things by connecting everyday objects.

Ubiquitous
Ubiquitous feature of IoT means the way of increasing the uses of IoT related device globally to allow world wide access and universal communication through makes its presence everywhere at once.

Heteroge
Heterogeneity characteristics of IoT allow a platform to communicate with a widespread variety of objects or devices through different network protocols like CoAP, MQTT. This key feature of IoT also allows the network connectivity among heterogeneous networks.

Dynamic Changes
The dynamic changes characteristics of Internet of Things allow the devices surrounding the system to change dynamically with any entity including location, person, temperature, time for collecting data from its surroundings.

Interoperability
Interoperability characteristics of Internet of Things support an wireless communication among various IoT objects using the same communication protocol.

Sensing and Actuation
Sensing and actuation is the key feature of IoT as the sensors are the backbone of the Internet of Things to measure and sense any variations in the surroundings to generate data for interacting with the IoT environment. Sensing mechanism of IoT provides the technology that alerts the people of the physical world, connected to this system.

Enormous Scale
This feature of Internet of Things connects a million of objects every day to communicate worldwide (Fig. 3).

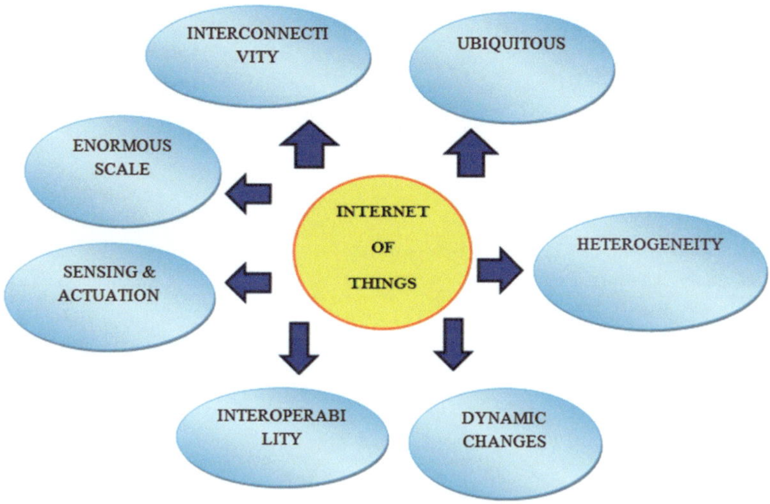

Fig. 3 Characteristics of Internet of Things

1.3 Need of IoT

Nowadays, it is difficult to explain the need of Internet Of Things. IoT is changing the meaning and dimension of our daily life by connecting our whole world. It is going to connect billions of daily life object. Internet of Thing means quick analysis with rapid sharing of enormous information.

To depict the need of Internet Of Things, just imagine a world where each and every devices in outside or inside world are linked in such a way that our home will invite us by unlocking the entrance door automatically for authorized members—lights and fans automatically turn on and off—all the electronics devices can be controlled from the outside.

1.4 Area of Applications

IoT invokes a technological revolution in Internet world with its numerous and diverse are of potential application. The new technology of IoT promises huge intelligent applications to connect the machine-to-machine, human-to-machine and an world-wide intelligent communication through the smart technologies like RFID, mobile computing, wireless sensor networks. The area of applications of IoT can be portrayed in Fig. 4.

Internet of Things is ruling the technical world for its various intelligent applications by making every object addressable, accessible and actionable. The IoT application serves a worldwide smart environment in all areas of everyday life through

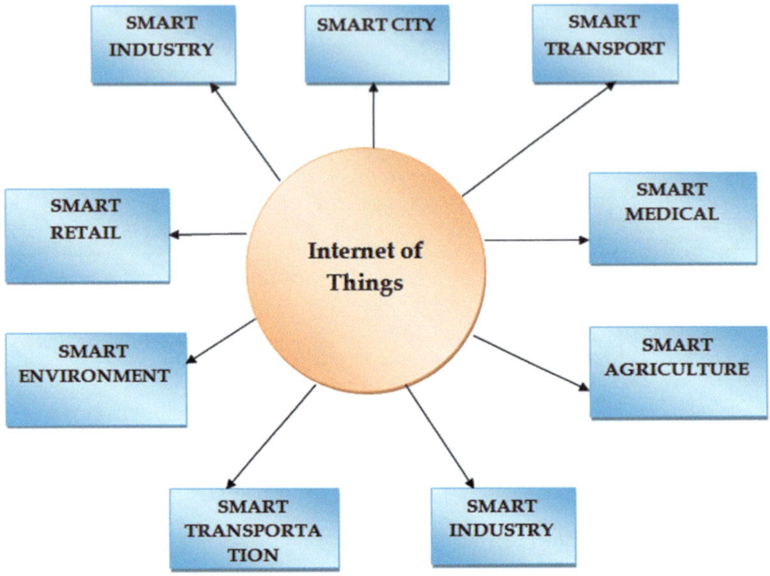

Fig. 4 Area of applications

the domains such as Life style, Transport system, Agriculture, Retail, Traffic system, Healthcare system, Farming system, Security system and so on.

1.4.1 IoT Applications in Smart City

It is expected the planet's inhabitants nearly 80% to subsist in metropolitan by 2040. This hasty urban enhance is already inserting a sizeable damage on the accessible communications. With additional citizens, it is making the shift to metropolitan livelihood and it's only departing to obtain inferior in the upcoming duration.

Cities are seeing exceptional growth, fetching key confronts as they seek to stay sustainable, vigorous and secure places for people to live and work. The IoT proposes new prospects for cities to cut pollution, make enhanced use of communications and remain people safe.

To assist cities make the most of the digital occasions, Nokia has launched "IoT for Smart Cities", a fully incorporated modular and scalable structure to proficiently deliver and manage elegant city services.

Benefits

- Modular sprightly scalable and safe
- Integrated smart city operations for effectiveness

- Swiftly launch innovative smart city services with secured performance and uniformity
- End to end fully incorporated approach

To provide accommodation this innovative requirement on metropolis about the planet is rotating to the Internet of Things novelty to improve their resources, trim down expenses and get better interaction. Although the probability is there for IoT to perk up nearly every part of metropolitan living, there are numerous IoT applications for smart cities.

A more resourceful Water Supply
IOT technique can also be a solution for water pollution and may be the reason for supplying the water in a scientific way. The IoT devices help in controlling the inflow and outflow data for leakage detection in a very efficient manner.

An Inventive Solution to Traffic Cramming
Another goal of IoT is giving the solution in the situation of road jam due to traffic cramming, which is a big problem for city life. The IoT devices like "Smart Traffic Signals" can be a solution for this problem by controlling the timings to show the right way to the vehicles and holiday traffics for keeping their movement properly.

The authorized city officials may collect the relevant data from different sources like road sensors, traffic cameras, vehicles, mobile phones for observing and detecting traffic hazards and incidents that happens in real time.

Apart from this, the transport drivers may be highly benefited through the IoT technique by getting the information regarding the accidents and road jam and can be alert for choosing the routes with less crowded accordingly.

Power-Efficient Buildings
Nowadays, the IoT technology acts as the base for the buildings with heritage communications to keep energy and improves their sustainability. The IoT approach is used by the building power organization systems to attach abnormal high temperature and fire protection systems to a central management application.

More Reliable Public Transportation
IoT techniques may help in the public shipping system when the system faces disturbances and interruptions at the time of rotten weather by assuring the transportation authority's the harmonized insights they need to execute contingency strategy, make sure that residents always have admittance to safe, reliable, and efficient. It could be done using imminent from cameras or associated devices at public areas.

Improved Public Safety
Smart cities often implement video supervising systems to begin the security concerns that turn up in every emergent city. The cities at this time have thousands of cameras observing traffic for accidents. Videotape investigative software also helps process the thousands of hours of video footage each camera produces, shaping it

down to only essential events. Systems using IoT technology revolve each camera fond of to the system into a sensor, with edge calculating and investigative starting right from the source. Then AI technology like machine learning will complete the investigation and send video recording to humans who can respond hastily to solve the troubles. Cities are also civilizing with elegant lighting ideas that replace conventional streetlights with associated LED communications. The LED lights last longer and conserve power and also supply information in real time.

1.4.2 IoT Applications in Agriculture

The exploit of Internet of Things in farming might contain the utmost shock, because the worldwide populace is located to touch 8.5 billion in the upcoming years. The agricultural production must embrace with IoT to nourish this too much population. There are so many threatens such as rising climate change, extreme bad weather conditions and momentous from exhaustive agricultural practices are creating challenges and therefore it is required to congregate the additional victuals.

Smart Cultivation
Elegant agricultural set on Internet of Things skill will facilitate cultivator to trim down ravage; it also improves production variety from the number of nourishment developed toward the amount of expedition the farmhouse vehicles have finished. Internet of Things supported elegant cultivation, a scheme is fabricated used for supervising the harvest field with the help of sensors like light, moisture, warmth, soil moisture and computerizing the cultivation system. The cultivator can observe the grassland situations as of everywhere.

Precision Farming
Precision agricultural be able to deliberation of everything which produces the farming performs further prohibited as well as exact when it move towards to elevating farm animals and increasing of harvest. The tactic of ranch organization, a crucial element is the make the most of different items like sensors, robotics, control systems, self-directed cars, computerized hardware, etc. The implementation of admittance to high-pitched Internet, transportable and electronics devices, and inexpensive satellites by the manufacturer are only some key technologies illustrating the accuracy cultivation development. The renowned applications of Internet of Things in farming area are Precision agriculture and many corporation is influencing this procedure throughout the globe. CropMetrics is a meticulousness farming corporation paying attention on current agronomic way out in the supervision of exactitude irrigation. The amenity of CropMetrics contains changeable Rate Irrigation (CRI) optimization, soil dampness surveys, effective optimizer PRO. CRI optimization exploits productivity on irrigated harvest grassland with soil changeability, gets better harvest, and enlarges water use competences. The soil wetness explore technology presents entire period local agronomy support and proposals to make the best water exploit effectiveness. The implicit optimizer PRO merges various skills for water managing into

a cloud supported planned for advisor and cultivator to take benefits in accuracy irrigation.

Agricultural Drones

Agricultural drones are superior paradigm of new technology in IoT. Cultivation is one of key productions to include drones. Drones are mainly utilized in cultivation to improve different farming procedures. The ways soil-supported and mid-air-based drones worn in cultivation are harvest health worn, harvest supervising, planting, harvest scattering, soil and ground investigation.

The key reimbursement of using agricultural drones contains incorporated GIS mapping that saves time and provide easiness of use to enlarge yields. Using drone policy and scheduling supported on authentic data collection and processing.

The farmers can draw insights from drone data about plant physical condition indices, plant counting, harvest prophecy, plant altitude dimension, field water positioning mapping, chlorophyll quantity, nitrogen contented in wheat, exhaust mapping, etc.

Livestock Monitoring

Vast farm proprietors are able to exploit wireless IoT appliances to gather data about the location and physical condition of livestocks. This information assists into recognizing farm animals which are ill so they could be alienated from the group, thus avoiding the widen of ailment. It can also minimize labor expenses as ranchers can set their farm animals with the assistance of Internet of Things-based sensors.

JMB is a reputed company of North America that proffers cow supervising way out to livestock manufacturers. The key solutions that facilitate the livestock owners examine cows that are expectant and regarding to give birth. As of the heifer, a sensor powered by battery is excluded when its water smashes. It forwards an information to the herd supervisor. During the occasion that is exhausted with heifers that are giving birth, the sensor allows farmers to be more alerts.

Elegant Greenhouses

The greenhouse agriculture is one of the most important methods to support the production of vegetables, flowers, fruits, crops, etc. in an attractive manner which deals with the ecological constraints through manual involvement that leads production loss, energy loss and huge labor cost. These inefficient approaches can be controlled by the implementation of IoT technology which may observe, monitor and also control the weather without any manual involvement.

In a smart greenhouse, various IoT-based sensors are used to measure the ecological constraints according to the plant condition. The system may use a cloud server, connected with IoT, for remotely accessing the system. In the greenhouse, the cloud server permits data processing and applies a control action.

Illuminum greenhouses is a drip mechanism which uses new skills for providing services by introducing new and reasonable greenhouses through lunar powered smart IoT sensors which monitors greenhouse condition and irrigates utilization through SMS aware to the farmer with an online gateway.

The IoT sensors provide the moistures, light stages and warmth in the greenhouse. These smart sensors also allow the actuators robotically to open and close the windows, switch on or off a fan or light within a Wi-Fi signal.

1.4.3 IoT Applications in SECURITY

All associated apparatus creates chances for assailant. These susceptible are wide, still for a single tiny device. The risks pretended comprise data transfer, device access and faulty devices. The main confront in security stays the security boundaries related to manufacturing stumpy cost devices, and mounting the number of devices which creates added chances for assaults.

There are many IoT applications for Internet of Things sanctuary:

Internet of Things system precautions: Defending and protecting the system involves Internet of Things tools to rear-end systems lying on the Internet. IoT network sanctuary is much difficult compare to usual network security as a lot of communication procedures and device potentialities, all of these pretense significant problems with enlarged complication. The key potential includes traditional endpoint protection characteristics like antivirus, antimalware, firewalls and interference deterrence and recognition systems.

IoT verification: As long as the facility for consumer to substantiate an Internet of Things device with managing numerous users of a single device like a connected car, ranging from easy fixed password to more healthy validation mechanisms such as digital certificates, biometrics and two-factor confirmation. Unlike a good number endeavor networks where the verification processes involve a human being inflowing a credential, a lot of IoT authentication circumstances such as embedded sensors are appliance-to-appliance based without any human involvement.

IoT encrypted: Encrypted data at rest and in also transfer among Internet of Things edging instruments and back-end systems uses a standard cryptographic algorithms serving and maintain data reliability and thwarting data snuffle from hackers. Internet of Things encryption should be convoyed by corresponding full encrypted key management processes, while deprived key management will cut overall security.

IoT in public key infrastructure: As long as entire X.509 digital certificate and cryptographic key and life-cycle potentials, including public and private key generation, distribution, management, and revocation. To exploit PKI the hardware specs for some Internet of Things instruments may bound and avert their capacity. Digitized certificates can be strongly encumbered onto Internet of Things devices on the time of assembling and then triggered by third-party software suites; the certificates might be mounted post-construction.

IoT sanctuary analytics: Collecting and normalizing data from Internet of Things devices and giving actionable coverage and alerting on specific activities. This elucidation is starting to add classy machine learning, artificial intelligence, and big data techniques to grant more predictive modeling and trim down the number of phony positives, but these capabilities are still emerging. Internet of Things security analytics will increasingly be required to detect IoT-specific assaults that are not recognized by traditional network security solutions such as firewalls.

IoT API sanctuary: As long as the capability to validate and approved data association between IoT tools and rear-end systems the API sanctuary will be necessary for defending the reliability of data transfer between rim devices and rear-end systems to make positive that only endorsed devices, developers, and applications are communicating with smart APIs as well as detecting potential threats and attacks against specific applications.

1.4.4 IoT Applications in Transport System

Roving with IoT

The IoT technology has a great impact on transport system. Day by day, the IoT devices are changing the concept of journey in worldwide by using the IoT-based technical concept through the sensors, mobile devices, etc.

The travel system in different modes become enriched with smart and intelligent technique of IoT application from starting of air tour to parking organization system.

Elegant Transportation and Traffic Management System

To utilize the modern and smart IoT technology, the transport system is ready to integrate the concept of IoT into the elegant roads for controlling the traffic hazards, road jamming and subsequent other traffic dangers. The IoT technology leads the concept of constructing the smart public roads with IoT sensors that would be efficient to detect the number of cars in each track and actuate the traffic lights accordingly, depending on this information. These sensors are also capable to transfer traffic in response to any disaster.

Though traffic as directed by the Internet of Things has numerous advantages, such as improving competence and dipping accidents by the exploit of actual-time monitoring, it can also lift the unease for probable cyber assaults. More improvement should be prepared for defending the inhabitants against such assault in the upcoming future if travel is to entirely incorporate by the IoT.

Parking

IoT concept is a very good choice for the city administrator for introducing an efficient and easier car parking system. An Internet of Things-based observing system has been incorporated in the District Department of Transportation's Park DC project in Washington, for alarming the drivers of available places by charging the appropriate charges for parking depending on the actual-time insist.

Using Sensors in Bridge

Internet of Things is also an essential option for different transportation plans, including bridges. IoT technique promises to introduce the modern system where the bridges would operate robotically, in the presence of boats, vehicles or any automobiles, with daily traffic system, without any manual interference.

Saving vast quantity of energy from road lights may possible by incorporating the IoT-based sensors in this system.

The drivers may be warned by the IoT sensors at the time of risk in frosty region on any others. The sensors also may inform the concerned persons if it is needed to repair or maintain the bridges.

Bicycle and Pedestrian Monitoring

Abetter, effective and scientific roadway plan may lead a danger-free and safe journey in busy time. To establish this concept in this modern and busy world, the application of Internet of Things may play an important role. It may improve the infrastructure and other features of the roadway plan for pedestrians and those roving by bicycle.

At this juncture, the IoT sensors may collect information regarding the method, by which several vehicles and pedestrians are on the road at any time, and the data which can be implemented into roadway plan.

Micro-navigation in Public Transportation

For giving some special and necessary facilities to the travelers of the city buses, the IoT may be an integral part of public transportation by applying the concept of "Macro-navigation" and "Micro-navigation". Both concepts may be incorporated into the IoS through smartphone applications.

Macro-navigation mentions to the responsibilities as significantly which bus will take the travelers to their exact location, whereas micro-navigation refers to the responsibilities of knowing which bus stop to depart at and which bus should be taken subsequently. The information for implementing macro- and micro-navigation in public transportation can be presented by text messages on a smartphone map, which can be downloaded by public transport traveler.

IoT in Airports and Air Travel

The airport authority of flight industry reuses the inexpensive and convenient Internet-connected sensors and various types of IoT tools of IoT-based technology in the development of airports and air travel to present them appropriately and effectively for the employees.

One of the burning examples of the application of IoT technology through the IoT smartphone applications in airport and air travel system is the "**MIA Airport Official**", a smartphone app, introduced by one of the challenging and popular airports in the USA, the Miami International Airport. The app can be downloaded on Android and iOS smartphone to provide different kinds of essential and important information to the travelers regarding travel and journey like location of airport eatery, luggage claim areas for introducing a very efficient, effective and modern system for both the travelers and air authority.

1.4.5 IoT in Medicine and Healthcare

The medicine & healthcare diligence are in a position of immense misery. Healthcare facilities are expenditure than ever, worldwide inhabitant is epoch, and the amount of unceasing syndrome is at a mount. We are forthcoming a globe everywhere in which the fundamental medical care would befit out of catch to the majority inhabitants and a vast division of civilization would move infertile due to aged people would be further level to persistent syndrome. Internet of Things application expansion is next to our salvage.

Although technology cannot prevent the inhabitants from eliminating unceasing diseases instantly, it can at least put up medical care simply on a compartment and in the matter of ease of access. Health investigative put away a huge portion of health center invoice. Technology can change the schedules of medicinal tests from a hospice to the sufferer's residence. The exact analysis will too decrease the necessity of medication.

A most recent hypothesis of IoT has a broad pertinent in several areas, as well as medical care. The complete principle of this prototype in medical care region is a combined anticipate since it all permits low medicinal centers to work proficiently and patients to get healthier treatment.

IoT in healthcare helps in:

- Tumbling disaster room waiting time
- Tracking patients and staff
- Enhancing drug management
- Ensuring accessibility of significant hardware

There are many benefits for healthcare and medicine:

Simultaneous Reporting and Monitoring
Real-time monitoring via connected devices can save lives in occurrence of a medicinal crisis such as liver problem, heart failure, kidney failure, brain stokes. With valid supervising of the situation in place by means of an elegant medical apparatus connected to a smartphone app, connected devices can collect medical and other necessary health data and use the data link of the smartphone to transfer composed information to a doctor.

The IoT mechanism accumulates and transfers health data such as blood pressure, blood sugar levels, weight, and ECGs. This information is accumulated in the cloud and can be distributed with an endorsed person, who might be a doctor.

End-to-End Connectivity and Affordability
IoT can lead an efficient and effective healthcare service through its interoperability characteristics by establishing a machine to machine communication for exchanging data. In this field, IoT provides a computerized patient care workflow system by healthcare mobility solution.

Connectivity Protocols
Different modern protocols like ZigBee, Wi-Fi, Z-Wave are capable to change the

method by which they speck illness and diseases in patients and also helps to serve better and modern technique of treatment.

Data assortment and Analysis
In modern healthcare system, it is necessary to transfer an enormous amount of information within a very small time because of their real-time applications. But it is difficult without the proper accessing of cloud as it leads an inefficient way to collect and manage of huge amount of information and also leads the hazards that can be a barrier to gain information from several sources for observing and testing physically.

At this juncture, the IoT tools may avoid to store the huge amount of raw data that it for psychoanalysis and report this data in real-time environment.

Tracking and Alerts
One of the main goals of IoT in medical field is the introduction of the devices that act like the alarming system which detects the critical information of a patient in a life-frightening conditions and transmit that information to the doctors for fast tracking and a speedy precaution accordingly. By transferring the reports, alarming messages and critical condition of the patient through the IoT-based apps and devices, irrespective of time and location, the IoT technology is becoming an integral part of our daily life.

This tracking and alerts method through the IoT-based technology allows a fast detecting, alerting, then tracing and monitoring of a patient in a crucial moment by doctors and also leads a way toward improved and perfect treatment.

Remote Medical Assistance
Medical facilities are still a daydream for some remote places of our country. In any emergency and critical situation, the patients of that remote location are not able to directly communicate with doctors due to distance. The IoT systems are introducing the mobility solutions for this problem in healthcare through which the experts may immediately examine the patients and can detect their problems. Apart from this, the IoT-based healthcare system also leads the initiatives that are forecasting to arrange the necessary medical apparatus and essential drugs on the basis of patient's requirement and disease-related information available via connected devices.

Nowadays, the promising contributions of IoT are making a big footprint on modern technical world due to its various uses in various aspects of our daily life. In the area of healthcare, the IoT has a lot of promises to overcome the difficulties and to bring a comfortable, easy and safe life.

Some of the equipment and wearable devices that depict the contribution of IoT system in healthcare are as follows:

Hearables
Hearables are the another application of IoT in medical field that changes the life of the people who are suffering from deafness by introducing the concept where this device may attach with Bluetooth for connecting smartphones with it that allows to balance, equalize and also include layered features to real-world sounds.

Ingestible Sensors
Another great wonder of IoT is the introduction of "Ingestible sensors". These sensors monitor and then give us warning message after detecting any abnormalities in our body. It is very much useful for the neuropatients and diabetic patients as these patients can be benefited by minimizing the symptoms and getting any odd feelings previously through the "Ingestible sensors".

Moodables
Another great achievement of IoT world is the concept of "Moodables"—the tools for enhancing our mood. Although it seems to be an imaginary concept but very soon this concept will be a real and great way to improve our mood throughout the day by applying the technique where moodables act like head-mounted wearable devices that send a low intensity of current toward our brain to raise our mood.

Computer Vision Technology
The drone technology is one of the great success of this modern technical world where the combination of AI and computer vision skill has promised to establish a technology that desire to replicate the visual insight and making review based on it (Fig. 5).

The Skydio Drones can take as an example of this technology which are used for computer visualization technology to detect and sense the abstracts and barriers for steer around them and also helps to guide visually damage people properly.

Healthcare Charting
Another successful contribution of IoT in healthcare system is "Audemix device" that helps to minimize much physical and manual work of doctors at the time of patient charting. This device captures and collects the patient's data for review in a comfortable way through powered by voice command for reducing doctor's work.

Fig. 5 Skydio Drones

1.4.6 IoT Applications in Retail System

In this smart world, the Internet of Things is ruling on the retail system by applying the methods of digital signage, antennas, smart tables and many more. With the help of IoT technology, the vendors are focusing on execution of their superior skills for making deep imminent into the customers desire and requirements for increasing sales.

Smart Shelves

Smart shelves are another witness of the footprint of IoT. It is the wireless inventory control systems that are fitted with weight sensors. The smart shelves are day by day establishing their necessity in the retail world. This system is containing the elements, namely RFID tag, RFID reader, transmitter. The RFID (radio frequency identification) tag is containing an integrated circuit and a computer chip antenna for transmitting information to the RFID reader which is collected from RFID tags.

Shelves are very much suitable for stock management that has been a comfortable process for a long time. The shelves can robotically observe information and forward alerts if any article is running low. The coupled devices are also crucial for avoiding inundate lack of goods and robbery in stores. By tracing records, they facilitate to trim down stress, eliminate operating faults and save expenses.

1.4.7 Robots in IoT

Robotics are the most attractive and amazing success of modern technology that are connected with IoT for using in retail world by presenting a widely transform the retail sector. Business Insider Intelligence are expecting that 2.7 million endeavor robots may dispatch between 2018 and 2022 as a lot of MNC companies have already incorporated the robots into their employment for minimizing the manual interference.

At this juncture, it can be mentioned that in San Francisco tested and experimented robots are used to help out the store shelves and also help to take the necessary record. In China, many of the restaurants already have introduced the expert robots in place of waiters. "Simbe Robotics" and "Fellow Robots" are well known for their very special and unique activities that help the shop visitors to quickly find their required products.

Beacons

Beacons are another effective and smart tools that connect Bluetooth for sending the notifications trustily to the smartphones of the shoppers' once they reach in the reporting area.

This technology allowed the customers to get information regarding different products, discounts, offers and many other information so that it making the customers more attract to go into the shop and make a buy of the products. Beacons are also acting as the helping hand of the Retail Corporation where they are used for the customer in-store routing, transferring messages and also collecting the information of client. Beacons also allows the retailers to enhance the customer reliability for improving sales by attaching the mobile applications with it. The technology of Beacon helps the vendors to bring the way which can communicate with their viewers for sharing their attractive offers very easily.

Digital Signature

Another strong, striking and a secured technique of this digital world is digital signature which already established their strong presence through the tools like Ipad screen, Touch screen, Instinctive touch screen. This method attracts the customers for rising the sales. By using the digital signage applications, the vendors are getting a lot of advantages by saving their costing related to the as usual print media through transferring the customer engagement toward next step.

Digital signage IoT solutions are enhancing the store skills owing to the Internet of Things. It also helps with the modification of prices of the advertise in real time.

2 Conclusion and Future Scope

The Interne of Things ensures to make a drastic change in everyday life for emerging a smart world. Its different applications in various fields are developing the world more digitalized and easier for us. It contains a key source of research due to its diverse application in numerous areas. This chapter will portray the key concept and necessity of IoT in our everyday life and also will reveal the impact of IoT in each and every areas of our life like medical, education, agriculture, transport, smart city, retail, environment.

Further Readings

1. Zhang, M., Qiu, Y., Zheng, R., Bai, X., Wei, W., Wu, Q.: A novel architecture for cognitive internet of things. Int. J. Secur. Its Appl. **9**(9), 235–252 (2015)
2. Manoj, T., Shetty, S.D., Trivarna,: Internet of things: the ultimate solution for the future. International journal of latest trends. In: Engineering and Technology Special Issue SACAIM 2017, pp. 539–544 (2017)
3. Sharma, V., Tiwari, R.: A review paper on "IOT" & "Its Smart Applications". Int J Sci, Eng Technol Res (IJSETR), 5(2) (2016)
4. Gubbi, J., Buyya, R., Marusic, S., Palaniswami, M.: Internet of things (IoT): a vision, architectural elements, and future directions. Future Gener. Comput. Syst. 29(7), 1645–1660 (2013)
5. Asseo, I., Johnson, M., Nilsson, B., Chalapathy, N., Costello, T.J.: The internet of things: Riding the wave in higher education. Educ. Rev., 11–31 (2016)

6. Cata, M.: Smart university, a new concept in the Internet of Things. In: The Proceedings of 14th RoEduNet International Conference—Networking in Education and Research (RoEduNet NER), pp. 195–197 (2015)
7. Sanju, D.D., Subramani, A., Solanki, V.K.: Smart city: IoT based prototype for parking monitoring & parking management system commanded by mobile app. In The Proceedings of Second International Conference on Research in Intelligent and Computing in Engineering, pp. 341–343. Poland (2017)

e-Device for the Protection of Agricultural Land from Elephant Attacks in Odisha: A Review

Sumit Badotra, S. N. Panda, K. S. Bath, Prasant Kumar Pattnaik, Rinkle Rani, Sarvesh Tanwar and Amit Sundas

Abstract Handling with wild animal's attacks is a special challenge for farmers throughout the world. Elephant attacks on agricultural land are one of the biggest social issues in some part of Odisha. They are confronting this issue when the yield is coming to dry and at the pre-collect stage. At that time the fragrance of harvests spread, by noticing the smell, an expansive gathering of elephants in the horticultural land wrecks the whole field and eventually harm crops by running over them, eating and totally vandalizing them. 100–120 elephants by and large lose their lives each year because of yield harm caused by them. Multiple traditional methods are still continued to use but these are both inefficient and insignificant. Therefore, to overcome the situation there has to be some system which can reduce the limitations of traditional methods used till now. So, here in this chapter we presented an approach to reduce crop damage by the elephants on agricultural lands. This proposed work is comprised of Internet of things and cloud-based technology with high-resolution camera to capture the images and process it and compare it with already stored images

S. Badotra (✉) · S. N. Panda · S. Tanwar · A. Sundas
Chitkara University Institute of Engineering and Technology, Punjab, India
e-mail: sumit.badotra@chitkara.edu.in

S. N. Panda
e-mail: snpanda@chitkara.edu.in

S. Tanwar
e-mail: sarvesh.tanwar@chitkara.edu.in

A. Sundas
e-mail: amit.sundas@chitkara.edu.in

K. S. Bath
Punjab State Council for Science and Technology, Chandigarh, India
e-mail: ksbath@yahoo.com

P. K. Pattnaik
KIIT University, Bhubaneshwar, India
e-mail: pratnaikprashant@gmail.com

R. Rani
Thapar University, Patiala, India
e-mail: raggarwal@thapar.edu

P. K. Pattnaik et al. (eds.), *IoT and Analytics for Agriculture*, Studies in Big Data 63,
https://doi.org/10.1007/978-981-13-9177-4_4

on cloud with the help of image processing algorithm and after detection it intimates the corresponding authority. This approach system would be fully automatic; it is a low-cost device and affordable for large area coverage and ultimately the lower the loss of lives.

Keywords Agricultural land · Crops · Animal attacks · Vandalizing · Internet of things · Cloud

1 Introduction

In India, every year around 150–200 individuals living in the edges of forest are slaughtered by elephants [1]. A large portion of the deaths occur in eastern Indian where people slaughter the elephants with burning tires, toxic substance and electric shock. In 2001, in India, 150 peoples were slaughtered in elephant attacks [1]. This was down from a pinnacle of around 300 individuals every year in the mid-1990s. Most of the people think that a larger number of individuals are killed because many attacks are in remote places and are not reported to government authorities. Wild animal attacks are becoming one of the serious concerns that need to be addressed because this not only loss of lives is increasing but it also affects the financial issues that are related to farmers as agriculture is one of the major earning sources to many farmers.

Importance of crops: One of the main requirements for livelihood of one's being is food [2]. The main source of our food product comes from mainly agricultural land [3]. India is an agrarian country, for the livelihood more than 60% of the population is dependent upon the agricultural land, and hence, it becomes mandatory to save the crops in agricultural field [4, 5]. All over in the India handling with wild animals is a big confrontation for farmers [6, 7]. Protection of their lands from wild animals is one of the special challenges for farmers throughout the world after drought. In search for food these animals like rabbits, moles, elephants, monkeys, dear, etc. either go alone or move in groups hence causes significant damage to crops [8].

They damage the crops in the agricultural land either by eating the plants or simply by running over the agricultural field [8]. Crop loss to farmers due to ravages by elephants in some states runs into millions of rupees each year. Therefore, these wild animals are becoming the one of the reasons for significant yield a loss that adds to the additional economic problems. Approximately 36% of the crop damage is done by these animals [8]. About 80% of the human population depends on the agriculture [8] and decade of forestry in Orissa 1981–1990 mentioned [9] "Amongst the tribal 61% engaged in cultivation, 21% as agricultural laborers, and 18% in forestry, mining, fishing and hunting, etc." [9]. We have targeted the agricultural land of farmers from Odisha those are affected by elephant's attack. The Scheduled Tribe (ST) population of Orissa was 22.21% (1991 census) of the total population, which is 22.13% in 2001 census. The State had the second largest tribal population

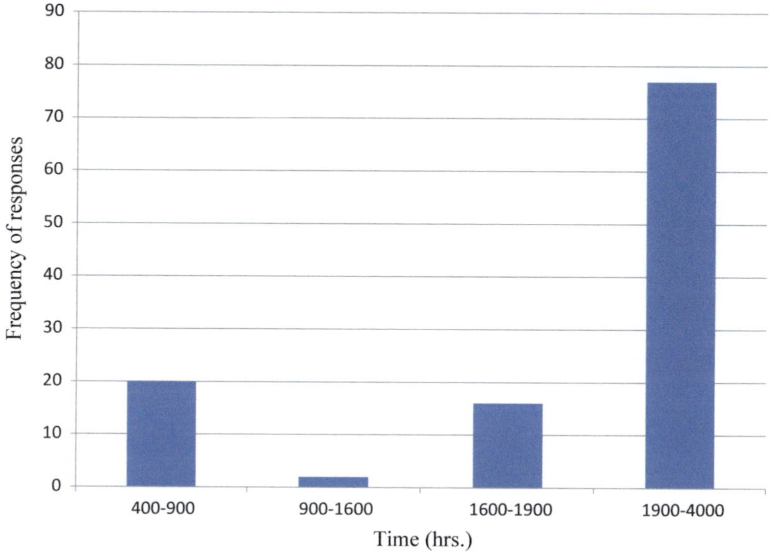

Fig. 1 Frequency of incursions by wild elephants into agricultural areas [12]

in India, next to Madhya Pradesh [10, 11]. Figure 1 shows the incursions by wild elephants into agricultural areas.

Crops in Odisha: Horticulture is an imperative division and a noteworthy marker of the state's financial status, farming remains the backbone of Odisha's economy and sustenance of the life of the general population. The normal harvests developed in the state are rice, beats, oil seeds, jute, roselle, sugarcane, coconut and turmeric which are critical products [2]. The satiate contributes one-tenth of the rice creation in India. Aside from nourishment grains, different money crops like jute, mesta, sugarcane, tobacco, elastic, tea, espresso and turmeric are likewise developed in the state. Odisha stands fourth in the creation of jute in the nation after West Bengal, Bihar and Assam. The agribusiness segment contributes just around 30% of the Gross State Domestic Product (GSDP), with over 60% population only bringing about low per capita pay in the ranch area [3]. The edited zone is around 87.46 lakh hectares out of which 18.79 lakh hectares are inundated. Atmosphere and soil assume a crucial part in Odisha's agribusiness economy [13]. In this manner, a noteworthy segment of the ranch arrive relies upon rain for water sustaining the product. The streams of Odisha are rain-nourished; ordinarily, it happens that the agriculturists must be reliant on the water system because significant parts of the generally developed land are being flooded. Additionally, the zone under farming is similarly more in the beach front regions of Odisha, i.e., Balasore, Bhadrak, Cuttack, Ganjam, Jajpur, Jagatasinghpur, Kendrapara, Khurda, Nayagarh, Puri [2, 3, 13].

The major crops in Odisha are as follows:

Rice: The crop is largely grown in the districts of Balasore, Cuttack, Mayurbhanj, Puri and Bhadrak (Fig. 2).

Fig. 2 Rice crop in Odisha [13]

Pulses: The most critical pulses developed in Odisha are gram, tuar, arhar and so on. Pulses are developed in the locale like Balasore, Bhadrak, Ganjam and Mayurbhanj in expansive overabundance.

Oil Seeds: The harvests like groundnut, mustard, castor and linseed are the principle oil seeds developed in Odisha. These are best created in the regions of Balasore, Kalahandi, Sambalpur and Dhenkanal [13].

Money Crops: Among money crops, fiber crops are particularly essential. Jute represents the most extreme zones. A lot of roselle is likewise developed in Odisha. Development of jute is essentially bound to the beach front fields of Cuttack, Balasore and Puri regions. It commands in the areas of Cuttack, Ganjam, Bolangir and Koraput locale. Sugarcane is the second most critical trade trim out Odisha in region and creation. It is developed in inundated zones. Odisha stands eight in sugarcane production in India. A lot of sugarcane is developed in Cuttack, Sambalpur, Bolangir, Kalahandi and Puri locale. Odisha additionally produces a little amount of tobacco. The new money trims, the most critical is cashew, which has been planted broadly in the low latiritic levels at the lower regions of the Eastern Ghats [13].

The Cashew Plantation Board is the in charge of the improvement of cashew development in the state. Cashew is additionally being planted on the sand-ridges on the whole seaside belt. This is a result of the reasonable edaphic and climatic conditions in these territories.

Elastic Plantations: To secure the debased slope slants of Eastern Ghats, elastic manor has been taken up by Rubber Board. Splendid prospect for the development and advancement of regular rubber exists in the northern Odisha Mayurbhanj, Keonjhar and Balasore because of good agro-climatic conditions. Odisha is a non-conventional region for elastic manors. Inland slope territories of Cuttack, Puri Ganjam and Dhenkanal locale have agro-climatic conditions reasonable for elastic development [13].

Cotton: Cotton development has been taken up broadly in Koraput, Bolangir and Kalahandi locale where reasonable soil and atmosphere conditions are found. Ocean Island cotton can be developed in the Ballipal area of Balasore region (Fig. 3).

Fig. 3 Vegetables in the agricultural land of Odisha [13]

Vegetables: The dirt and climatic states of Odisha are good to grow an assortment of vegetables consistently. Real vegetables developed in the state are brinjal, tomato, onion, cole yields, okra and gourds [13].

Elephant attacks in Odisha: The problem of crop depredation and loss of lives due to elephants is also severe in eastern region of India (also known as Central India—i.e., South West Bengal, Jharkhand and Orissa) [9]. This paper aims only in presenting Orissa state from India as an affected area for protecting agricultural lands from elephant attacks. In India among the wild animals, elephants are responsible for maximum crop damage and so subjected to retaliatory killing by the farmers for protection of crop and property. Incidents such as damaging the crops and destroying them are happening in the nearby or adjacent areas to the forests resulting in emergence of a direct conflict between humans and animals [14–16]. Crops such as paddy, millets, maize sugarcane and banana are the most affected in India. The loss of human lives is very serious in India, where on an average 200 people are being killed every year by wild elephants [17], and a substantial number of these incidents happen on cultivated land by raiding elephants or elephants raiding harvested stored grains. Unfortunately, the age-old conflict [17] between elephant and people has shifted from the use of space (landscape) to resources like property and life [9, 17]. In the process, not only elephant populations have been affected, but the people have also suffered through depredation of cultivated crops and loss of live. Therefore, it is becoming the necessity of an hour to prevent and save the crops from these attacks. Figure 5 depicts how elephants harm the crops (Fig. 4).

Fig. 4 Elephant attack in agriculture land [18]

Fig. 5 An Asian elephant calf and its mother flee from a mob, throwing fiery balls of tar, in the Indian state of West Bengal [16]

Affected crops: Of the 1700 villages sampled, 1249 villages were affected by elephant depredation (during the time frame of the sampling, i.e., from 1992 to the sampling date of the individual sampled forest division) [6]. The most vulnerable crop was paddy in 1126 (90%) affected villages, of these, on the periphery 681 villages, within forests 198 villages and in the 'rest' 247 villages. Sugarcane was most vulnerable in 22 (2%) affected villages (periphery—16, within forest—5, and in the rest—1) [6]. Jackfruit was most vulnerable in 14 (1%) affected villages (periphery—7, and within forest—7) [6]. Maize and banana were most vulnerable in 8 affected villages each (0.6%). Vegetables were most vulnerable in 4 affected villages (0.3%). Palm tree, ragi (madia), coconut plants, and various fruit plants were most vulnerable in one or two affected villages. Apart from these above-mentioned crops, various other crops were also affected by elephant depredation [6]. These were wheat, kulthi (horse-gram), biri (black gram), green gram, sweet-potato, date-palm trees, mango fruits, bamboo, cashew plants (root), sesame, groundnut and stored paddy-straw. In this state, paddy is the most vulnerable crop that needs protection from elephant ravages [6] which is cultivated in more than 72% of the cropped area. In more than 90% affected villages, this crop was targeted by the elephants [6].

Elephant-Man Conflict: There has been colossal rising in the man-elephant conflicts in Odisha which is obvious from the way that 82 people were executed specifically elephant attacks till February 2018 [19]. 66 individuals had kicked the can in elephant strikes in 2016–2017 while 82 people were butchered, 55 others hurt in 2017–2018 [19]. Hordes of wild elephants have destroyed 1119 houses and standing yields in around 5401.98 areas of place where there is arrive in different parts of the state [17, 19]. Also, 143 wild creatures including 11 elephants were murdered by poachers amid the most recent four years in the state [9, 15, 20]. It is declared that the lawmaking body has fail to characterize any strong framework to shield the elephants from blasting into human normal surroundings which fundamentally prompts such man-animal conflicts [9, 15, 20]. The Asian elephants (*Elephas maximus*) is exceptionally debilitated by territory discontinuity, living space misfortune and human-elephant conflict [20].

India has 60% of Asian elephant masses, around 66% of the elephants live either close or inside human-directed scenes [9]. Southern India harbors half of India's elephant people containing around 6300 elephants [15]. The expansion in human populace in India moved by rural and mechanical development has prompted the transformation of the woods lands into human settlements [17]. Because of this, the wild elephant and other creature populaces confront intense deficiency of assets, for example, water and sustenance, influencing them to move frequently into the human environment. Henceforth, there has been serious man-elephant struggle [17]. The contention has been on the ascent in the backwoods outskirt zones with crowds of wild pachyderms straying into human home [9]. It is affirmed that the lawmaking body has fail to characterize any strong framework to shield the elephants from blasting into human regular surroundings which fundamentally prompts such man-animal conflicts [21]. The Asian elephants (*E. maximus*) are incredibly crippled by an area irregularity, living space disaster and human-elephant difficulty [7, 22].

The dazzling photograph as shown in Fig. 6 of an Asian elephant calf on fire has won a characteristic life photography contention—pulling in overall thought with respect to the battle regions of elephant-human conflict in India [16].

In 2014–2015, the team has recorded 277 incidents of yield attacks, 12 occurrences of property harm, three human wounds and one human demise in 17 towns [23]. In such attacks 200 elephants were killed in the vicinity of 2006 and 2011 across India [18]. While crop-raiding happened consistently, it topped directly after the rainstorm; elephants assaulted for the most part palatable crops, for example, coconut, areca nut and banana. Typically, villages that mutual longer limits with forests were more in danger than others. Crop damage is the most major issue of conflict in the zone.

The Department of Wildlife Sri Lanka Conservation appraises that an aggregate of 1163 elephant lost their lives in the wild in the vicinity of 1950 and 1970, of which 639 (55%) were murdered by agriculturists with regard to their yields [20]. An aggregate of 452 elephant passing's was accounted for between the mid-1980s to mid-1990 in the north western and focal areas alone, of which 336 (or 74%) were

Fig. 6 Agricultural fences [26]

executed by farmers [21]. There are no legitimate records of passing's of agriculturists in connection to elephant trim attacking or yield insurance practices of ranchers. De Silva reports that all things considered 30–50 individuals are slaughtered by wild elephants every year in Sri Lanka [20]. As indicated by Santiapillai [8] seven years before, more than 500 individuals have lost their lives because of the contention among people and elephants. In India, in 2000, 100 individuals and 30 elephants were executed in occurrences including a solitary crowd of 300 elephants. Table 1 shows the loss of life in men versus elephant [8].

Earlier preventive measures used to avoid elephant attacks: 547 people have been killed in Odisha between April 2010 and December 2017. 288 have been harmed in encounters with wild elephants. At a similar period, 571 elephants have passed on [1, 24]. Villagers had tried multiple attempts to prevent their crops from the elephant attacks. The various traditional methods which are used by them are as follows:

- *Crackers*: Crackers were used to make animals afraid hence stopping them to come in the agricultural lands.
- *Shouting*: This was the manual method; people make noises and shout so that animals listen to them and in fear of human beings they run away.
- *Use of Noise-Making Instruments*: In this method people from the affected area make use of various noise-making instruments with the help of such tools noise is created to make animals move from the agricultural/village area.
- *Man-Days/Hours Used in Guarding Crop*: During the time interval in which animals come, men from the affected area used to guard crops for hours. In this method there was a risk of their lives (both human and animal) because any animal attack can happen during this time which ultimately leads to death and injury only.
- *Agricultural Fences*: Fencing is a well-known wild creature security practice that can keep going for a long time for wild animal protection. However, using wall as training is frequently controlled. Some neighborhood and state elements may

Table 1 Loss of life in men versus elephant

State	Year	Men		Elephant
		Injured	Died	
Orissa	1990–2000	35	86	
Bihar	1980–1991	134		
Bangladesh	2001	40		
	2002	10	3	
Sri Lanka	Every year	30	50	100–120
Karnataka	2006–2007	16		
	2008–2009	46		
	2009–2010	41		
Maharashtra	2003–2003	21	10	
West Bengal	2008–2009	273	94	
Assam	1994–2004	250		230

confine or keep the utilization of specific sorts of wall. In this manner, before choosing a reasonable fence, it is essential to check neighborhood law directions. The nature of fencing relies upon the material and structure. The most commonly used fences are plastic, wire and electric fences shown in Fig. 6 [25, 26].
- *Natural anti-agents*: A couple of agriculturists lean toward using regular safety efforts instead of mechanical or substance defensive practices. There are diverse ways to deal with shield crops from wild creatures, for example, egg-based anti-agents, stew peppers, smoke, castor oil and bee sanctuaries [26].
- *Biophysical obstructions*: These are minimal effort hone which are made of coconut tree group, bamboo sticks or shrubs [25, 26]. But these strategies are might not totally powerful to ensure the products against wild creatures.
- *Chemical anti-agents*: Chemicals like methyl anthranilate and butanethiol can likewise be utilized to shield edit from wild creatures [26].
- *Electronic anti-agents*: The powerful, dependable and eco-accommodating procedure without harming creatures is electronic anti-agents, for example, ultrasonic—high recurrence sound (Hooter) and sonic—discernible commotion [25, 26].

The aforementioned methods are not effective and uses the manual approach only, to overcome the situation and various losses; here, we are proposing a solution that shall be containing the capabilities. This work is comprised of Internet of Things (IoT) and cloud-based technology with high-resolution camera to capture the images in real-time mode and processes with the help of image processing algorithm. After the processing, it detects the arrival of elephants and immediately switches on the gadgets automatically and wirelessly, it also forwards all this to the concerned authority in their smartphones. It would be a low-cost device and affordable for large area coverage. This method is better than available solutions because in the traditional methods used by the farmers they themselves had to provide protection to their farmlands with their constant manual supervision

Rest of the chapter is divided into the following sections. Section 2 is describing the related literature survey, Sect. 3 is comprised of the problem statement for protecting the agricultural lands from the wild animal attacks, Sect. 4 contains the proposed approach to limit the disadvantages of the current techniques used for the protection of lands from wild animal attacks, Sect. 5 involves the methodology used to implement the proposed approach, and finally Sect. 6 is comprised of conclusion with future scope.

2 Literature Survey

Agriculture is the primary occupation for over 75% of people in Odisha [1]. Out of which 85% are small and marginal farmers with one acre to three acres of land-holdings. In [27] the authors exhibited the few purposes behind product attacking, including the expanded annihilation and fracture of environments because of formative ventures. Authors in [28], defense strategies are the moves which are made by ranchers to deflect natural life from assaulting crops. Adapting techniques are particular moves made by ranchers to beat the impacts of yield striking and in addition to maintain a strategic distance from trim attacks on their farms. Automated animal checking with vitality productive remote gadgets mounted on the creatures can enhance effectiveness of cultivating industry and increment its benefit by diminishing dependence on human work [12]. We talk about the commonsense arrangement of defer tolerant store and forward actions observing the design that gives information maintenance, distinguishing custom occasions, warning issuing, remote and in situ questions replying. The center of this engineering, a novel vitality proficient, interruption tolerant, Mobile Ad Hoc Network (MANET) steering convention (EERD), gives offloading information to long haul stockpiling by sending information to cultivate servers by means of sinks that are a piece of a MANET and handles in situ questions issued by clients gathered with the creatures. The vitality proficiency is accomplished by powerful adjustment to the present conduct of the creatures conveying the observing gadgets. The current recognition frameworks, for example, Light and Camera Option, infra stable and Seismic Communication Option are profoundly costly and less successful.

Drawbacks of existing technologies: The existing techniques for the prevention of wild animals entering into the agricultural field are to develop electric barriers, crackers & forced throat water on the boundaries of agricultural area. These techniques will be harmful for the wild animals like elephant, pigs, monkeys, etc.

Numerous different techniques are likewise used to maintain a strategic distance from the elephant assaults and the contention too. Development of elephant verification trenches is being done everywhere throughout the world. Pradhan et al. [24] examined sun-based fencing to stay away from elephant-human clash. In [29] they exhibited the idea of utilizing bee sanctuaries to alleviate elephant trim theft. In [30] authors talked about the part of the counterfeit water sources which enable elephants to dwell in timberlands amid dry seasons. In [31], the creators examined the potential utilization of satellite innovation for strife alleviation. The elephants labeled with radio collars respond viciously and harm it and even the elephants pass on. In [32] authors proposed the likelihood of utilizing the elephant-elephant correspondence (elephant thunders) to recognize the nearness of a group of elephants in closeness. In this work, the creators have recorded the low recurrence infrasound design; however, they do not contrast with that of different creatures with affirmation of an elephant event. In [33] they proposed unmanned flying machine framework to review elephants, in which the elephant pictures are gained at a stature of 100 m; however, the little flight time and being costly do not make it reasonable. In [34], authors proposed appearance-based acknowledgment calculations for recognizable proof of elephants.

Given the frontal face picture of an elephant, the framework looks through the individual elephant utilizing vision calculations and gives the outcome as officially distinguished elephant, or as another recognizable proof. In [35] authors exhibit an elephant photograph ID framework in light of the shape examination of the scratches describing the elephant's ears. In [27] tended to recognizing elephants from photos, and looking at resultant catch recover based populace parameter gauges utilizing directed visual distinguishing proof of individual varieties in tusk, ear overlap and projection shape. The creators demonstrate that this is a solid strategy for singular distinguishing proof and consequent estimation of populace parameters. Yet, progressively, the catch of elephant's front picture is not conceivable (Fig. 7).

Authors [36] examined the potential utilization of satellite innovation for struggle relief. The elephant GPS radio neckline transmits the geo areas to the satellite to screen the development of the groups in the timberland locales as appeared in Fig. 8. The mind-boggling expenses of as far as possible its utilization to close to few creatures. The elephants labeled with radio collars respond savagely and harm it. A 20-year-old male elephant was shot with a sedated dash by forest and WWF specialists for settling the radio neckline. In the ebb and flow explore, dealing with wild creatures a savvy-based hindrance assurance framework is wanting to create, which has innocuous contrasted with the before strategies. In the investigation of wild animals, it lights up that with a couple of sounds wild animals will exasperate and escape from that place.

Authors [17] talked about electric sunlight-based fencing to stay away from elephant-human clash. Electric wall is seen by nearby individuals as a changeless arrangement and there is an impulse to introduce fences wherever there is strife as appeared in Fig. 8.

It is less demanding to pursue elephants before they enter fields and consequently most harm can be turned away [37]. Guarding from watch towers, watching, and excursion wire cautions furnish agriculturists with guidance ahead of time of moving toward elephants [38]. Once the creatures are identified, dynamic yield guarding gadgets utilizing light and commotion are sent to pursue them away. An early caution-

Fig. 7 Electriccollar [36]

Fig. 8 Electric fencing [17]

ing framework to limit the human-elephant struggle in the woodland outskirt zones is proposed in this chapter. The framework mitigates such clashes in two different ways: (i) giving cautioning to individuals about the foreseen section of elephants into human habitation; (ii) giving development data to the experts to make a move to pursue the pachyderms back to the forest.

3 Problem Statement

To limit the elephant attacks on agricultural lands, there has to be some approach or method through which it can detect the elephant attack on agricultural land and suitable actions are to be taken in concern of that.

The main issues with traditional approach that is used to protect lands from wild animal attacks include the following challenges:

- Damage of individuals
- Loss of human life
- Decimation of farm infrastructure
- Destruction of crops
- Formation of a domain of fear.

4 Proposed Approach

The proposed solution shall be containing the capabilities which ultimately overcome the limitations of traditional methods. This work is comprised of IoT and cloud-based technology with high-resolution camera to capture the images in real-time mode and process with the help of image processing algorithm. After the processing it detects the arrival of elephants and immediately switches on the gadgets automatically and wirelessly, it also forwards all this to the concerned authority in their smartphones. It would be a low-cost device and affordable for large area coverage.

This method is better than available solutions because in the traditional methods used by the farmers they themselves had to provide protection to their farmlands with their constant manual supervision.

The overall objective of the proposed approach are as follows:

1. Protect farms from animals such as elephants.
2. Design of low-cost device to send real-time information to the farmers.
3. To help the forest department to acknowledge them (Fig. 9).

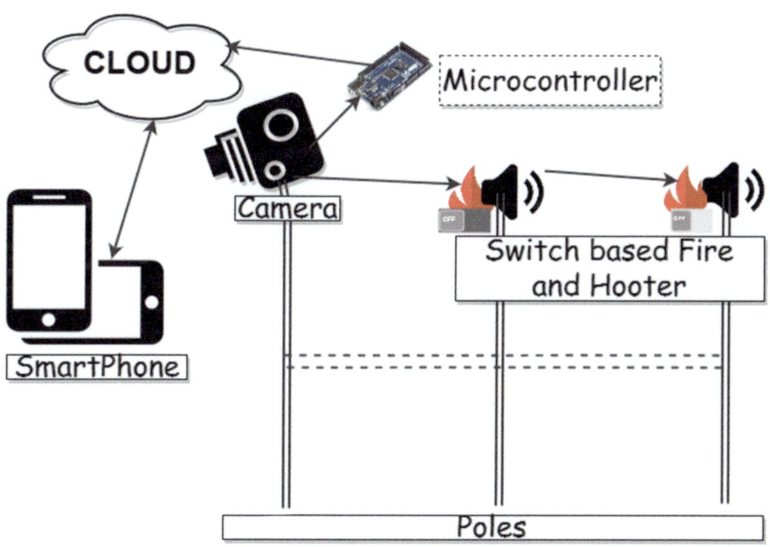

Fig. 9 Block diagram of the proposed approach

5 Methodology

a. The aim of the proposed mechanism is to detect the elephant using image processing on microcontroller, store that image on cloud storage and send an alert via application to the authorized person through smartphone.
b. The capturing/sample of image compared using image processing system.
c. This automatic elephant detection system can capture the image which is used for identifying or comparing the images in the database. Camera module is interfacing to the microcontroller. It is used to capture an image and send captured image to the microcontroller. When an image is captured by camera, it is taken by microcontroller to compare with the database image.
d. After comparing images output is positive/negative then it gives commands to application used in smartphone. Application is used to send a message to the authorities and the concerned persons. After comparison output is either positive or negative. If output is positive then elephant image is detected and vice versa.
e. The audio unit consists of a microcontroller, audio player, memory to store tracks and the speakers. The microcontroller will be programmed such that when the camera detects an animal and send an image.
f. Smartphone application used to send the message to the owner of the farm and the nearest forest department if the animals cross the specified limit.

This framework presents a technique to automate the way toward keeping off the wild animals from farmlands and furthermore give reconnaissance differentiating among the authorized and unauthorized individuals. Figure 10 describes the overall working of the proposed approach

- Installation of camera on the pole, incorporated with microcontroller.
- Installation of switch-based artificial fire & hooter on nearby poles.
- Database will be maintained on cloud.
- Cloud will be attached with smartphone app to capture the real-time scenario.

Applications

- Protection of farmer life.
- This can also be implemented in forest range for preventing the animal reaching the road.
- It can be used as security system for houses.
- This unit can be used in area where human movement is minimal. It can be placed near agricultural lands to prevent the entry of wild animals from nearby forest.
- This can also be implemented in forest ranges for preventing the animals reaching the roads (Fig. 11).

Advantages

- Continuous monitoring can be done
- Mini appearance.
- Device can be implemented in large scale.

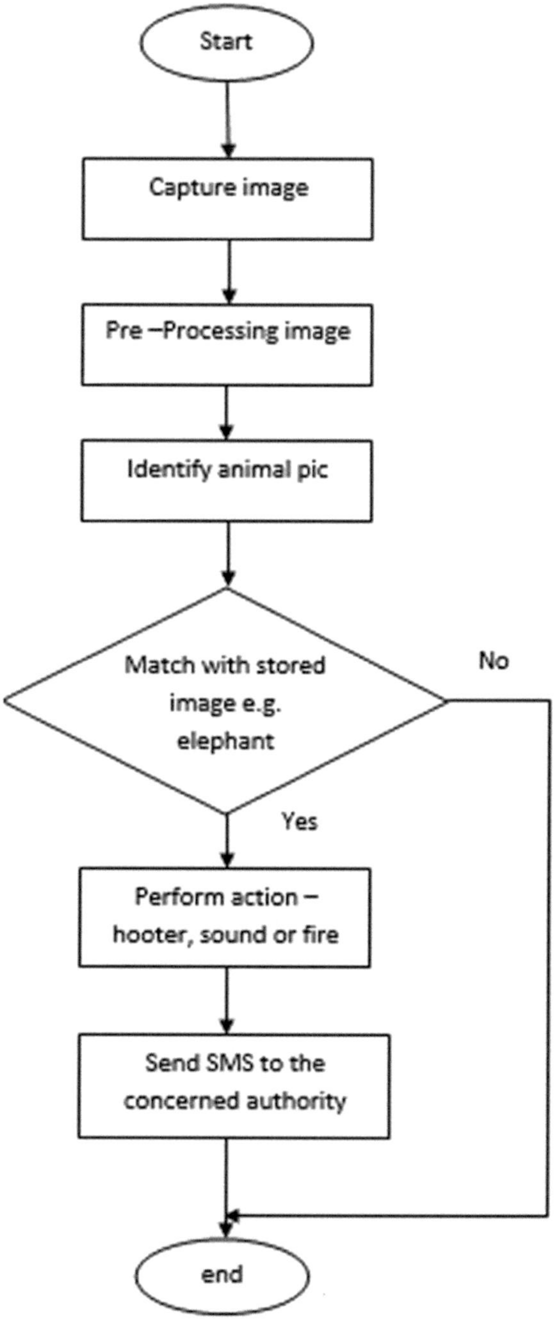

Fig. 10 Flowchart of the proposed approach

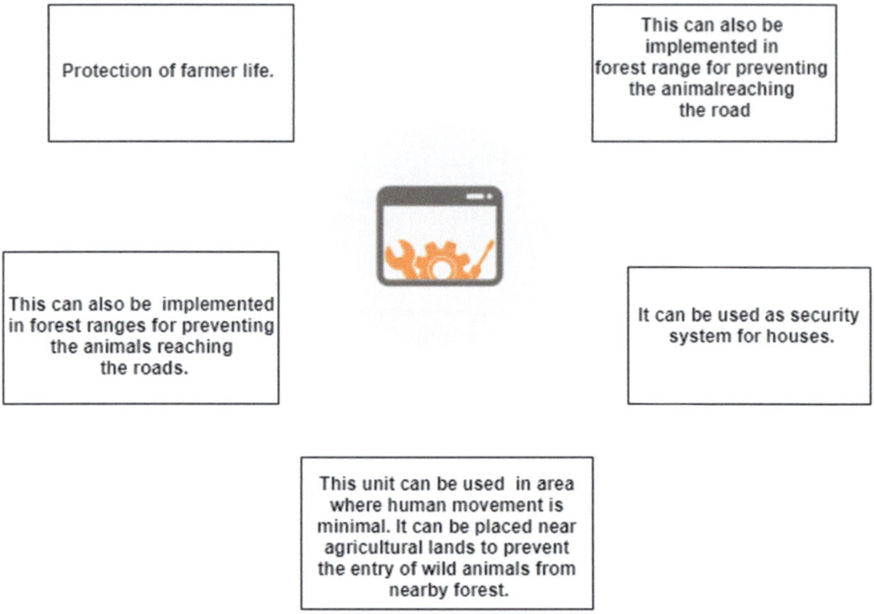

Fig. 11 Applications of e-Device

- Simple circuit.
- Low-cost maintenance.
- Automatic operation.
- Long life time.
- Real-time scenario/protocol.
- Highly flexible and economical system for crop protection.
- Less Human Effort (Ultimately will reduce number of deaths of both human and elephants).
- It saves time and energy as it does not require human supervision.
- Solar charging- low power requirement.
- Database handling.
- Netis connectivity can be achieved.
- Use of latest technology—Technical flexibility can be achieved like—drone can be used for surveillance.

6 Conclusion

Protection of crops against wild animals is an important task with advantages to the farmers and society as a whole. The main objective of this chapter is to provide an effective solution to this problem, so that the economic losses incurred by our farmers are minimized and they have a good crop yield. It helps to save time and money by reducing the manual work that is otherwise required if the farmers themselves had to provide protection to their farmlands with their constant manual supervision. This proposed solution is comprised of IoT and cloud and with the help of high-resolution camera, elephant coming toward the agricultural land can be detected and recognized using some image processing algorithm and simultaneously directs the sound gadgets and artificial fire to start. Through the smartphone application concerned authorities and owners of the land are intimated about the same. This system is environment friendly and needs no source of external power supply as it can work with solar energy.

References

1. https://www.thehindu.com/todays-paper/tp-national/tp-otherstates/jumbo-problem-for-odisha-forest-officials/article22217568.ece. Accessed 1st July 2018
2. Sulek, J.M., Hensley, R.L.: The relative importance of food, atmosphere, and fairness of wait: the case of a full-service restaurant. Cornell Hotel Restaurant Adm. Q. **45**(3), 235–247 (2014)
3. McMichael, A.J., Powles, J.W., Butler, C.D., Uauy, R.: Food, livestock production, energy, climate change, and health. The Lancet **370**(9594), 1253–1263 (2007)
4. Laha, A.K.: Interval-valued data analysis. Int. J. Bus. Analytics Intell. **3**(2) (2015)
5. Praveen, K.V., Suresh, A., Reddy, A.A., Singh, D.R.: Risks and adaptation strategies in rainfed agriculture in india: an analysis. Indian J. Agric. Sci. **88**(6), 958–963 (2018)
6. Bindu, D., Dilip kumar, M.D., Mamtha, B., Prashanth, P.: Prevention of wild animals entering into the agriculture fields. Int. J. Eng. Basic Sci. Manage. Soc. Stud. **1**(1), 639–647 (2017)
7. Goswami, V.R., Lauretta, M.V., Madhusudan, M.D., Karanth, K.U.: Optimizing individual identification and survey, effort for photographic capture-recapture sampling of species with temporally variable morphological traits. Anim. Conserv. **15**(2), 174–183 (2012)
8. Santiapillai, C., Wijeyamohan, S., Bandara, G., Athurupana, R., Dissanayake, N., Read, B.: An assessment of the human-elephant conflict in Sri Lanka. Ceylon J. Sci. (Biological Sciences) **39**(1), 44–52 (2010)
9. Chanchal Kumar Sar: Project: Elephant-Human conflict in Asia report on Orissa–India (Part-I) Liz Claiborne/Art Ortenberg Found. Technical Report, Aug 2009
10. http://censusindia.gov.in/Tables_Published/SCST/dh_st_madhya_pradesh.pdf. Accessed 26th June 2018
11. Raghavendra, K.J., Suresh, A.: Risk management in rainfed agriculture: an analysis of strategies and adaptations followed by farmers in Madhya Pradesh. Indian J. Agric. Sci. **88**(6), 895–901 (2018)
12. Nakandala, M.S., Namasivayam, S.S., Chandima, D.P., Udawatta, L.: Detecting wild elephants via WSN for early warning system. In: 2014 7th International Conference on Information and Automation for Sustainability (ICIAFS), IEEE, pp. 1–6 (2014)
13. http://creativeodisha.com/GeneralInformation/Agriculture-in-Odisha. Accessed 31st July 2018
14. Perera, B.M.A.O.: The Human-Elephant conflict: a review of current status and mitigation methods. Gajah: J. IUCN/SSC Asian Elephant Specialist Group **30**, 41–45 (2009)
15. Pooja, G., Bagali, M.U.: A smart farmland using Raspberry Pi crop vandalization prevention & intrusion detection system. Int. J. Adv. Res. Innovative Ideas Educ. (IJARIIE). **1**(5) (2016). ISSN (O)-2395-4396
16. https://www.nationalgeographic.com/photography/proof/2017/11/elephants-flee-fire-in-award-winning-photo/. Accessed 24th June 2018

17. Doyle, S., Groo, M., Sampson, C., Songer, M., Jones, M., Leimgruber, P.: Human-Elephant Conflict—What Can We Learn from the News? (2005)
18. https://thewire.in/environment/elephants-conflicts-trenches-beehive-yellapur-corridor. Accessed 30th June 2018
19. https://www.thehindu.com/news/national/other-states/elephant-attacks-in-odisha-district-kill-6/article23631072.ece. Accessed 26th June 2018
20. Bandara, R., Tisdell, C.: Asian elephants as agricultural pests: economics of control and compensation in Sri Lanka. Nat. Resour. J., 491–519 (2002)
21. Meredith, R.: The *Elephant* and *The Dragon*: The Rise of India and China and What it Means for All of Us, 6th edn., (2007)
22. Mamboleo, A.A., Doscher, C., Paterson, A.: Are Elephants the most disastrous agricultural pests or the agents of ecological restorations? J. Biodivers. Endangered Species **5**(1), 1–10 (2017). ISSN: 2332-2543
23. https://www.thehindu.com/news/national/kerala/the-elephant-in-my-farm/article24301218.ece. Accessed 30th June 2018
24. Pradhan, R.N., Behera, P., Mohapatra, R.K., Sahu, H.K.: Human-Elephant conflict in Dhenkanal Forest Division, Odisha, India. Afr. J. Basic Appl. Sci. **5**(4), 193–199 (2013)
25. Mamo, D., Bouer, H., Tesfay, Y.: Crop damage by African elephant's assessment in Kaftasheraro National Park, Ethiopia. Afr. J. Ecol. **52**(2), 138–143 (2014)
26. Das, S.K., Chattopadhyay, S.: Human fatalities from wild elephant attacks-a study of fourteen cases. J. Forensic Legal Med. **18**(4), 154–157 (2011)
27. Ardovini, A., Cinque, L., Sangineto, E.: Identifying elephant photos by multi-curve matching. Pattern Recogn. **41**(6), 1867–1877 (2008)
28. Bakels, J.: Farming the forest edge: perceptions of wildlife among the Kerinci of Sumatra. In: Knight, J. (ed.) Wildlife in Asia: Cultural Perspectives, pp. 147–164. Outledge-Curzon, Taylor & Francis Group, London and New York
29. Fernando, P., Kumar, M.A., Williams, A.C., Wikramanayake, E., Aziz, T., Singh, S.M.: Review of Human-Elephant conflict mitigation measures practiced in South Asia, AREAS Technical Support Document Submitted to World Bank. WWF-World Wide Fund for Nature, Vaud, Switzerland (2008)
30. King, L.E., Lawrence, A., Douglas-Hamilton, I., Vollrath, F.: Beehive fence deters crop-raiding elephants. Afr. J. Ecol. **47**(2), 131–137 (2009)
31. Loarie, S.R., Aarde, R.J.V., Pimm, S.L.: Fences and artificial water affect African savannah elephant movement patterns. Biol. Conserv. **142**(12), 3086–3098 (2009)
32. Venkataraman, A.B., Saandeep, R., Baskaran, N., Roy, M., Madhivanan, A., Sukumar, R.: Using satellite telemetry to mitigate elephant-human conflict: an experiment in northern West Bengal, India. Curr. Sci. **88**(11), 1827–1831 (2005)
33. Venter, P.J., Hanekom, J.J.: Automatic detection of African elephant (Loxodontaafricana) infrasonic vocalisations from recordings. Biosyst. Eng. **106**(3), 286–294 (2010)
34. Vermeulen, C., Lejeune, P., Lisein, J., Sawadogo, P., Bouche, P.: Unmanned aerial survey of elephants. PLoS ONE **8**(2), 1–7 (2013)
35. Dabarera, R., Rodrigo, R.: Vision based elephant recognition for management and conservation. In: Proceedings of the 5th International Conference on Information and Automation for Sustainability (ICIAfs '10), pp. 163–166, Dec 2010
36. Arun, B., Venkataraman, R., Sandeep, N., Baskaran, M., Roy, A., Madhivanan, Sukumar, R.: Using satellite telemetry to mitigate Elephant–Human conflict: an experiment in Northern West Bengal, India. Curr. Sci. **88**(11), 1827–1831 (2005)
37. Lenin, J., Sukumar, R.: Action Plan for the Mitigation of Elephant-Human Conflict in India. Final Report to The U.S. Fish and Wildlife Service. Asian Nature Conservation Foundation, Bangalore, India (2011)
38. Santhiya, S., et al.: A smart farmland using Raspberry Pi crop prevention and animal intrusion detection system. Int. Res. J. Eng. Technol. (2018)

The Impact of Irrigation on Agricultural Productivity in the Bolpur Sub Division, West Bengal

Subhasis Mondal

Abstract Bolpur Sub Division is located at the south-eastern part of the Birbhum district, West Bengal. This study area is intensively and extensively irrigated by means of Canal, Submersible pump, Tank, and River lift. By virtue of irrigation practice, the entire part of this Sub Division has been successfully brought under cultivation. Above all, the modern socioeconomic parameters comprehensively give rise to a higher degree of agricultural development, where irrigation plays the role of a catalyst. Due to the presence of irrigation, almost all the farms have more than 90% of their land as the NSA (net sown area). In the study area, monsoons are active only for 3–4 months in a year. The remaining 8–9 months are the dry season. The crops want irrigation for successful growth in the dry season. The ever increasing population leads to more intensive agriculture which needs more irrigation facilities, along with other inputs. Agricultural productivity of any region is closely associated with a number of physical (like relief, climate, soil, water, etc.), socioeconomic, political, institutional, and organizational factors of that region. A higher level application of chemical fertilizers and pesticides, timely and adequate supply of irrigation, and higher level supervision help to have an upper level productivity of crops in the study area.

Keywords Irrigation · Agricultural productivity

1 Introduction

There is an important role of irrigation in the economy, depended on agriculture. The total quantity of agricultural land in the Bolpur Sub Division is 218,900 acre. Almost, those 135,937 acre under the facility of irrigation, i.e. 62.1% of the total agricultural land is irrigated. This study area is intensively and extensively irrigated by means of Canal, Submersible pump, Tank, and River lift sources of irrigation. By virtue of irrigation practice, the entire part of this Sub Division has been successfully brought under cultivation. Above all, the modern socioeconomic parameters comprehensively

S. Mondal (✉)
The Department of Geography, Visva-Bharati University, Bolpur, West Bengal, India
e-mail: rasasubhasismondal@gmail.com

P. K. Pattnaik et al. (eds.), *IoT and Analytics for Agriculture*, Studies in Big Data 63,
https://doi.org/10.1007/978-981-13-9177-4_5

give rise to a higher degree of agricultural development, where irrigation plays the role of a catalyst. Due to the presence of irrigation, almost all the farms have more than 90% of their land as the NSA. Four blocks of the Sub Division depend on agriculture. Crops are produced twice in a year, almost 83,188 acre land. As the chief crops in Bolpur Sub Division Rice, Potato, Mustard seed, Til (*Sesamum orientale*) and Mursuri (*Ervum hirsutum*) may be mentioned. Recently, the production of Boro cultivation has increased due to extension of irrigation practices.

2 Statement of Problems

- In the study area, monsoons are active only for 3–4 months in a year. The remaining 8–9 months are the dry season. The crops want irrigation for successful growth in the dry season.
- There are certain crops such as Rice, Potato and Mustard seed which require more water. Rice is the principal crop of the study area.

3 Literature Survey and Review

Literature survey is the documentation of a wide-ranging review of the published and unpublished work from secondary sources of data in the area of specific interest, to the researcher. The researcher has studied several secondary and primary researches from several books, journals, and documents for information on the research topic. Reviewing the literature on the topic area at that time, helped the researcher a lot, to focus further on interviews more meaningfully, on certain aspects which were found to be important in published studies. The measurement of production and inputs required for the production of that output is known as agricultural productivity. In other words, it is an input–output ratio. Several studies have been made on this issue, some of which are discussed below.

Todkari [5] studied "Impact of irrigation on agriculture productivity in the Sholapur district of Maharashtra state." He found that irrigation is a decisive factor for agricultural development. Inadequacy of rainfall calls for irrigation for successful agriculture.

Motebennur [4] studied "Impact of irrigation on agricultural development in the Dharwad district of Karnataka state." He found that irrigation is very important input of agriculture. Without irrigation very little can be expected from extensive cultivation. Agriculture without irrigation in areas having less than fifteen centimeters rainfall is a suicidal uneconomic venture.

4 Objectives of the Study

- To find out the impact of irrigation on productivity of crops.
- To uncover the problems and suggest policy measure.

5 Database

The data have been collected from the two sources, i.e.,

- Secondary, which is collected from different books, references, journals, and different offices and
- Primary, which is collected through field investigation.

6 Methodology

Methodology is very vital part on which the quality of research highly depends. Keeping in mind the objectives of this research, five (5) villages namely Ruppur and Benuria from Bolpur-Sriniketan Block, Goltikuri from Illambazar Block, Kurumba from Labhpur Block, and Saota from Nanoor Block were purposively selected as a sample village for a survey. Every sample village is 100% irrigated. Thus, the total number of Blocks was four (4). Finally, a total number of two hundred fifty (250) households were surveyed (fifty (50) households from each sample village).I computed the interrelationship between the independent and dependent variable by Co-relation technique. I also computed the cause and affect relationship by regression analysis.

GIS map: The location map has been prepared through the application of ARC GIS 10. This section also deals with the whole process of obtaining and processing of data. This study is mainly based on the primary data; however, some secondary data have also been used.

7 Location of the Study Area

Bolpur Sub Division is located at the south-eastern part of the Birbhum district, West Bengal. The total area of the Bolpur Sub Division is 1275.93 km^2. It extended from 23° 32′ 30″ to 23° 53′ 00″ north latitudes and from 87° 23′ 30″ to 87° 57′ 30″ east longitudes. This Sub Division comprises four blocks viz. Illambazar in the west (261.54 km^2 area), Bolpur- Sriniketan in the middle (334.58 km^2 area), Nanoor Block in the east (311.83 km^2 area), Labhpur in the north (367.98 km^2 area). Ajoy

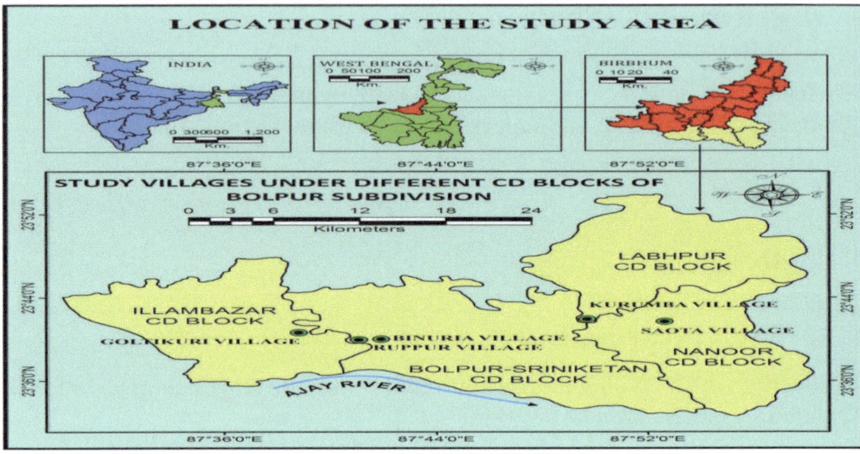

Fig. 1 Showing location map of Bolpur Sub Division

river in the south detaches Bolpur Sub Division from Burdwan district. Keeping in mind the objectives of this research, five (5) villages namely Ruppur and Benuria from Bolpur-Sriniketan CD Block, Goltikuri from Illambazar CD Block, Kurumba from Labhpur CD Block, Saota from Nanoor CD Block were purposively selected as sample village for survey. Every sample village is 100% irrigated (Fig. 1).

8 Impact of Irrigation on Productivity of Crops

The measurement of production and inputs required for the production of that output is known agricultural productivity [3]. It is actually an input output ratio. Agricultural productivity of any region is closely associated with a number of physical (like relief, climate, soil, water, etc.), socioeconomic, political, institutional and organizational factors of that region. Here, in the following section, the productivity of different crops in terms of their output (weight) is discussed one by one.

8.1 Productivity of Aman Paddy

Aman paddy, being the major food grain of the area, is cultivated by all the farmers in the study villages. The study shows that the farmers in this area produced this crop at the rate of 7.95–25.10 quintals per acre of land, but majority of them produces as low as in between 13.67 and 19.38 quintals per acre. Higher levels of productivity of Aman paddy are observed in Kurumba, Saota, and Goltikuri villages, followed by Binuria and Ruppur villages. Those who have percentage of NIA (net irrigated area) to NSA (net sown area) are most of big and medium farmers. Big and medium farmers can buy Submersible water, seeds, and fertilizers easily and they produce more.

8.1.1 Correlation Analysis Between Percentage of NIA to NSA and Productivity of Aman Paddy in Quintal Per Acre in the Study Villages

Positive correlation coefficient between the percentage of NIA to NSA and agricultural productivity of Aman paddy in quintal per acre has been observed in all the sample villages (Saota, Kurumba, Goltikuri, Binuria and Ruppur), which means that both the variables are directly proportional to each other. The computed t values of all the sample villages (Saota, Kurumba, Goltikuri, Binuria, and Ruppur) are significant at 1 and 5% level of significance. Therefore, the null hypothesis is rejected and the alternative hypothesis is accepted indicating that there is significant statistical correlationship between the percentage of NIA to NSA (X variable) and agricultural productivity of Aman paddy in quintal per acre (Y variable) (Table 1).

8.1.2 Regression Analysis Between Percentage of NIA to NSA and Agricultural Productivity of Aman Paddy in Quintal Per Acre in the Study Villages

In the case of **Saota** village, the regression coefficient (b) is (0.57) where the equation is ($Y = -33.03 + 0.57x$). This shows that per unit increase of percentage of NIA to NSA will increase 2.90 unit of agricultural productivity of Aman paddy in quintal per acre in this village. The coefficient of determination (r^2) is 0.97, which means 97% of the total variation in Y (agricultural productivity of Aman paddy in quintal per acre) is being explained by X (percentage of NIA to NSA). The intercept (a) is -33.03 which means the regression coefficient shows variation below this constant figure.

In the case of **Kurumba** village, the regression coefficient (b) is (0.56) where the equation is ($Y = -32.01 + 0.56x$). This shows that per unit increase of percentage of NIA to NSA will increase 0.56 unit of agricultural productivity of Aman paddy in quintal per acre in this village. The coefficient of determination (r^2) is 0.94, which means 94% of the total variation in Y (agricultural productivity of Aman paddy in quintal per acre) is being explained by X (percentage of NIA to NSA). The intercept (a) is -32.01 which means the regression coefficient shows variation below this constant figure.

In the case of **Goltikuri** village, the regression coefficient (b) is (0.48) where the equation is ($Y = -25.05 + 0.48x$). This shows that per unit increase of percentage of NIA to NSA will increase 0.48 unit of agricultural productivity of Aman paddy in quintal per acre in this village. The coefficient of determination (r^2) is 0.99, which means 99% of the total variation in Y (agricultural productivity of Aman paddy in

Table 1 Correlation analysis between percentage of NIA to NSA and productivity of Aman paddy in quintal per acre in the study villages

Villages	Independent variable (X)	Dependent variable (Y)	r value	Computed t value	Tabulated value		Level of significance	Null hypothesis	Alternative hypothesis
					1%	5%			
	Percentage of NIA to NSA	Agricultural productivity of Aman paddy in quintal per acre							
Saota			0.98	40.47	2.66	2.00	1 and 5%	Rejected	Accepted
Kurumba			0.97	27.72	2.66	2.00	1 and 5%	Rejected	Accepted
Goltikuri			0.99	69.56	2.66	2.00	1 and 5%	Rejected	Accepted
Binuria			0.99	100.01	2.66	2.00	1 and 5%	Rejected	Accepted
Ruppur			0.99	100.16	2.66	2.00	1 and 5%	Rejected	Accepted

Degree of freedom (df) = 48
Source Computed by the researcher

Table 2 Regression analysis between percentage of NIA to NSA and agricultural productivity of aman paddy in quintal per acre in the study villages

Villages	Independent variable (X)	Dependent variable (Y)	b (slope)	a (intercept)	R^2
	Percentage of NIA to NSA	Agricultural productivity of Aman paddy in quintal per acre			
Saota			0.57	−33.03	0.97
Kurumba			0.56	−32.01	0.94
Goltikuri			0.48	−25.05	0.99
Binuria			0.39	−16.69	0.99
Ruppur			0.50	−29.69	0.99

Source Computed by the researcher

quintal per acre) is being explained by X (percentage of NIA to NSA). The intercept (a) is −25.05 which means the regression coefficient shows variation below this constant figure.

In the case of **Binuria** village, the regression coefficient (b) is (0.39) where the equation is ($Y = -16.69 + 0.39x$). This shows that per unit increase of percentage of NIA to NSA will increase 0.39 unit of agricultural productivity of Aman paddy in quintal per acre in this village. The coefficient of determination (r^2) is 0.99, which means 99% of the total variation in Y (agricultural productivity of Aman paddy in quintal per acre) is being explained by X (percentage of NIA to NSA). The intercept (a) is −16.69 which means the regression coefficient shows variation below this constant figure.

In the case of **Ruppur** village, the regression coefficient (b) is (0.50) where the equation is ($Y = -29.69 + 0.50x$). This shows that per unit increase of percentage of NIA to NSA will increase 0.50 unit of agricultural productivity of Aman paddy in quintal per acre in this village. The coefficient of determination (r^2) is 0.99, which means 99% of the total variation in Y (agricultural productivity of Aman paddy in quintal per acre) is being explained by X (percentage of NIA to NSA). The intercept (a) is −29.69 which means the regression coefficient shows variation below this constant figure (Table 2).

8.2 Productivity of Boro Paddy

Boro paddy is the second most highly productive food crop, produced in the study area. Higher level application of chemical fertilizers and pesticides, timely and adequate supply of irrigation, and higher level supervision help to have an upper level productivity of this crop. Generally, the productivity ranges here in between 12 and 28.50 quintals per acre in different parts of the study villages. Kurumba and Binuria villages obtain better production from per acre of land by cultivating this crop. On the other hand, farmers of Saota and Goltikuri villages produce this crop with moderate amount of productivity, whereas Ruppur village has low level of productivity of this crop.

8.2.1 Correlation Analysis Between Percentage of NIA to NSA and Productivity of Boro Paddy in Quintal Per Acre in the Study Villages

Positive correlation coefficient between the percentage of NIA to NSA and agricultural productivity of Boro paddy in quintal per acre has been observed in all the sample villages (Saota, Kurumba, Goltikuri, Binuria, and Ruppur), which means that both the variables are directly proportional to each other. The computed t values of all the sample villages (Saota, Kurumba, Goltikuri, Binuria, and Ruppur) are significant at 1 and 5% level of significance. Therefore, the null hypothesis is rejected and the alternative hypothesis is accepted indicating there is significant statistical correlationship between the percentage of NIA to NSA (X variable) and agricultural productivity of Boro paddy in quintal per acre (Y variable) (Table 3).

8.2.2 Regression Analysis Between Percentage of NIA to NSA and Agricultural Productivity of Boro Paddy in Quintal Per Acre in the Study Villages

In the case of **Saota** village, the regression coefficient (b) is (0.57) where the equation is ($Y = -29.30 + 0.57x$). This shows that per unit increase of percentage of NIA to NSA will increase 0.57 unit of agricultural productivity of Boro paddy in quintal per acre in this village. The coefficient of determination (r^2) is 0.98, which means 98% of the total variation in Y (agricultural productivity of Boro paddy in quintal per acre) is being explained by X (percentage of NIA to NSA). The intercept (a) is -29.30 which means the regression coefficient shows variation below this constant figure.

In the case of **Kurumba** village, the regression coefficient (b) is (0.54) where the equation is ($Y = -26.50 + 0.54x$). This shows that per unit increase of percentage of NIA to NSA will increase 0.54 unit of agricultural productivity of Boro paddy in quintal per acre in this village. The coefficient of determination (r^2) is 0.95, which means 95% of the total variation in Y (agricultural productivity of Boro paddy in quintal per acre) is being explained by X (percentage of NIA to NSA). The intercept (a) is -26.50 which means the regression coefficient shows variation below this constant figure.

In the case of **Goltikuri** village, the regression coefficient (b) is (0.45) where the equation is ($Y = -18.51 + 0.48x$). This shows that per unit increase of percentage of NIA to NSA will increase 0.45 unit of agricultural productivity of Boro paddy in

Table 3 Correlation analysis between percentage of NIA to NSA and productivity of Boro paddy in quintal per acre in the study villages

Villages	Independent variable (*X*)	Dependent variable (*Y*)	*r* value	Computed *t* value	Tabulated value		Level of significance	Null hypothesis	Alternative hypothesis
	Percentage of NIA to NSA	Agricultural productivity of Boro paddy in quintal per acre			1%	5%			
Saota			0.99	49.42	2.66	2.00	1 and 5%	Rejected	Accepted
Kurumba			0.97	30.40	2.66	2.00	1 and 5%	Rejected	Accepted
Goltikuri			0.99	51.36	2.66	2.00	1 and 5%	Rejected	Accepted
Binuria			0.99	104.26	2.66	2.00	1 and 5%	Rejected	Accepted
Ruppur			0.98	41.81	2.66	2.00	1 and 5%	Rejected	Accepted

Degree of freedom (df) = 48
Source Computed by the researcher

Table 4 Regression analysis between percentage of NIA to NSA and agricultural productivity of Boro paddy in quintal per acre in the study villages

Villages	Independent variable (*X*)	Dependent variable (*Y*)	*b* (slope)	*a* (intercept)	R^2
	Percentage of NIA to NSA	Agricultural productivity of Boro paddy quintal per acre			
Saota			0.57	−29.30	0.98
Kurumba			0.54	−26.50	0.95
Goltikuri			0.45	−18.51	0.98
Binuria			0.30	−4.33	0.99
Ruppur			0.49	−23.94	0.97

Source Computed by the researcher

quintal per acre in this village. The coefficient of determination (r^2) is 0.98, which means 98% of the total variation in *Y* (agricultural productivity of Boro paddy in quintal per acre) is being explained by *X* (percentage of NIA to NSA). The intercept (*a*) is −18.51 which means the regression coefficient shows variation below this constant figure.

In the case of **Binuria** village, the regression coefficient (*b*) is (0.30) where the equation is ($Y = -4.33 + 0.30x$). This shows that per unit increase of percentage of NIA to NSA will increase 0.30 unit of agricultural productivity of Boro paddy in quintal per acre in this village. The coefficient of determination (r^2) is 0.99, which means 99% of the total variation in *Y* (agricultural productivity of Boro paddy in quintal per acre) is being explained by *X* (percentage of NIA to NSA). The intercept (*a*) is −4.33 which means the regression coefficient shows variation below this constant figure.

In the case of **Ruppur** village, the regression coefficient (*b*) is (0.49) where the equation is ($Y = -23.94 + 0.49x$). This shows that per unit increase of percentage of NIA to NSA will increase 0.49 unit of agricultural productivity of Boro paddy in quintal per acre in this village. The coefficient of determination (r^2) is 0.97, which means 97% of the total variation in *Y* (agricultural productivity of Boro paddy in quintal per acre) is being explained by *X* (percentage of NIA to NSA). The intercept (*a*) is −23.94 which means the regression coefficient shows variation below this constant figure (Table 4).

8.3 *Productivity of Potato*

The Potato is largely cultivated in the sample villages. Farmers here, produce 50–100 quintals of Potato from per acre of land. In terms of higher levels of productivities of the farmers have been observed in Kurumba, Saota, Goltikuri, and Binuria villages.

Here, in those villages, controlled supply of Tank irrigation and Submersible pump irrigation, higher level application of chemical fertilizers and pesticides, use of HYV seeds, experience of the farmers, etc., help to produce this crop with larger amount of productivity per acre of land. At Ruppur village, the farmers also produce potato with moderate amount of productivities. The study also shows that the farmers of Ruppur village are least interested to cultivate this crop due to soil condition.

8.3.1 Correlation Analysis Between Percentage of NIA to NSA and Productivity of Potato in Quintal Per Acre in the Study Villages

Positive correlation coefficient between the percentage of NIA to NSA and agricultural productivity of Potato in quintal per acre has been observed in all the sample villages (Saota, Kurumba, Goltikuri, Binuria, and Ruppur), which means that both the variables are directly proportional to each other. The computed t values of all the sample villages (Saota, kurumba, Goltikuri, Binuria, and Ruppur) are significant at 1 and 5% level of significance. Therefore, the null hypothesis is rejected and the alternative hypothesis is accepted indicating there is significant statistical correlationship between the percentage of NIA to NSA (*X* variable) and agricultural productivity of Potato in quintal per acre (*Y* variable) (Table 5).

8.3.2 Regression Analysis Between Percentage of NIA to NSA and Agricultural Productivity of Potato in Quintal Per Acre in the Study Villages

In the case of **Saota** village, the regression coefficient (b) is (3.31) where the equation is ($Y = -208.62 + 3.31x$). This shows that per unit increase of percentage of NIA to NSA will increase 3.31 unit of agricultural productivity of Potato in quintal per acre in this village. The coefficient of determination (r^2) is 0.97, which means 97% of the total variation in Y (agricultural productivity of Potato in quintal per acre) is being explained by X (percentage of NIA to NSA). The intercept (a) is −208.62 which means the regression coefficient shows variation below this constant figure.

In the case of **Kurumba** village, the regression coefficient (b) is (3.14) where the equation is ($Y = -203.24 + 3.14x$). This shows that per unit increase of percentage of NIA to NSA will increase 3.14 unit of agricultural productivity of Potato in quintal per acre in this village. The coefficient of determination (r^2) is 0.95, which means 95% of the total variation in Y (agricultural productivity of Potato in quintal per acre) is being explained by X (percentage of NIA to NSA). The intercept (a) is −203.24 which means the regression coefficient shows variation below this constant figure.

In the case of **Goltikuri** village, the regression coefficient (b) is (2.51) where the equation is ($Y = -138.27 + 2.51x$). This shows that per unit increase of percentage of NIA to NSA will increase 2.51 unit of agricultural productivity of Potato in quintal per acre in this village. The coefficient of determination (r^2) is 0.99, which means

Table 5 Correlation analysis between percentage of NIA to NSA and productivity of Potato in quintal per acre in the study villages

Villages	Independent variable (X)	Dependent variable (Y)	r value	Computed t value	Tabulated value		Level of significance	Null hypothesis	Alternative hypothesis
					1%	5%			
	Percentage of NIA to NSA	Agricultural productivity of Potato in quintal per acre							
Saota			0.98	18.17	3.12	2.20	1 and 5%	Rejected	Accepted
Kurumba			0.97	16.27	3.01	2.16	1 and 5%	Rejected	Accepted
Goltikuri			0.99	215.17	3.25	2.26	1 and 5%	Rejected	Accepted
Binuria			0.98	19.10	3.01	2.16	1 and 5%	Rejected	Accepted
Ruppur			0.98	15.82	3.36	2.31	1 and 5%	Rejected	Accepted

Source Computed by the researcher

Table 6 Regression analysis between percentage of NIA to NSA and agricultural productivity of Potato in quintal per acre in the study villages

Villages	Independent variable (X)	Dependent variable (Y)	b (slope)	a (intercept)	R^2
	Percentage of NIA to NSA	Agricultural productivity of Potato in quintal per acre			
Saota			3.21	−208.62	0.97
Kurumba			3.14	−203.24	0.95
Goltikuri			2.51	−138.27	0.99
Binuria			3.18	−190.30	0.97
Ruppur			3.20	−191.91	0.97

Source Computed by the researcher

99% of the total variation in Y (agricultural productivity of Potato in quintal per acre) is being explained by X (percentage of NIA to NSA). The intercept (a) is −138.27 which means the regression coefficient shows variation below this constant figure.

In the case of **Binuria** village, the regression coefficient (b) is (3.18) where the equation is ($Y = -190.30 + 3.18x$). This shows that per unit increase of percentage of NIA to NSA will increase 3.18 unit of agricultural productivity of Potato in quintal per acre in this village. The coefficient of determination (r^2) is 0.97, which means 97% of the total variation in Y (agricultural productivity of Potato in quintal per acre) is being explained by X (percentage of NIA to NSA). The intercept (a) is −190.30 which means the regression coefficient shows variation below this constant figure.

In the case of **Ruppur** village, the regression coefficient (b) is (3.20) where the equation is ($Y = -191.91 + 3.20x$). This shows that per unit increase of percentage of NIA to NSA will increase 3.20 unit of agricultural productivity of Potato in quintal per acre in this village. The coefficient of determination (r^2) is 0.97, which means 97% of the total variation in Y (agricultural productivity of Potato in quintal per acre) is being explained by X (percentage of NIA to NSA). The intercept (a) is −191.91 which means the regression coefficient shows variation below this constant figure (Table 6).

8.4 Productivity of Mustard Seed

Mustard seed is the most important oil seed produced in this area, and so it is largely cultivated in all the sample villages. But the productivity of this Rabi crop is moderate, ranging in between 8 and 11.11 quintals per acre land. Majority of farmers produce it with an average amount of productivity. In this group, the farms of Goltikuri village predominate numerically followed by Saota, Kurumba, Ruppur, and Binuria villages. But the cultivators of Goltikuri village produce it with larger amount per acre, in such farmers use larger amount of NPK and Urea fertilizer and provide necessary and adequate irrigation.

8.4.1 Correlation Analysis Between Percentage of NIA to NSA and Productivity of Mustard Seed in Quintal Per Acre in the Study Villages

Positive correlation coefficient between the percentage of NIA to NSA and agricultural productivity of Mustard seed in quintal per acre has been observed in all the sample villages (Saota, Kurumba, Goltikuri, Binuria, and Ruppur), which means that both the variables are directly proportional to each other. The computed t values of all the sample villages (Saota, Kurumba, Goltikuri, Binuria, and Ruppur) are significant at 1 and 5% level of significance. Therefore, the null hypothesis is rejected and the alternative hypothesis is accepted indicating there is significant statistical correlationship between the percentage of NIA to NSA (X variable) and agricultural productivity of Mustard seed in quintal per acre (Y variable) (Table 7).

8.4.2 Regression Analysis Between Percentage of NIA to NSA and Agricultural Productivity of Mustard Seed in Quintal Per Acre in the Study Villages

In the case of **Saota** village, the regression coefficient (b) is (0.21) where the equation is ($Y = -8.56 + 0.21x$). This shows that per unit increase of percentage of NIA to NSA will increase 0.21 unit of agricultural productivity of Mustard seed in quintal per acre in this village. The coefficient of determination (r^2) is 0.99, which means 99% of the total variation in Y (agricultural productivity of Mustard seed in quintal per acre) is being explained by X (percentage of NIA to NSA). The intercept (a) is -8.56 which means the regression coefficient shows variation below this constant figure.

In the case of **Kurumba** village, the regression coefficient (b) is (0.21) where the equation is ($Y = -8.58 + 0.21x$). This shows that per unit increase of percentage of NIA to NSA will increase 0.21 unit of agricultural productivity of Mustard seed in quintal per acre in this village. The coefficient of determination (r^2) is 0.99, which means 99% of the total variation in Y (agricultural productivity of Mustard seed in quintal per acre) is being explained by X (percentage of NIA to NSA). The intercept (a) is -8.58 which means the regression coefficient shows variation below this constant figure.

In the case of **Goltikuri** village, the regression coefficient (b) is (0.16) where the equation is ($Y = -3.86 + 0.16x$). This shows that per unit increase o f percentage of NIA to NSA will increase 0.16 unit of agricultural productivity of Mustard seed in quintal per acre in this village. The coefficient of determination (r^2) is 0.98, which means 98% of the total variation in Y (agricultural productivity of Mustard seed in quintal per acre) is being explained by X (percentage of NIA to NSA). The

Table 7 Correlation analysis between percentage of NIA to NSA and productivity of Mustard seed in quintal per acre in the study villages

Villages	Independent variable (*X*)	Dependent variable (*Y*)	*r* value	Computed *t* value	Tabulated value		Level of significance	Null hypothesis	Alternative hypothesis
					1%	5%			
	Percentage of NIA to NSA	Agricultural productivity of Mustard seed in quintal per acre							
Saota			0.99	135.16	2.92	2.12	1 and 5%	Rejected	Accepted
Kurumba			0.99	112.35	2.95	2.13	1 and 5%	Rejected	Accepted
Goltikuri			0.98	26.54	2.88	2.10	1 and 5%	Rejected	Accepted
Binuria			0.99	98.38	2.98	2.15	1 and 5%	Rejected	Accepted
Ruppur			0.99	84.59	2.95	2.13	1 and 5%	Rejected	Accepted

Source Computed by the researcher

Table 8 Regression analysis between percentage of NIA to NSA and agricultural productivity of Mustard seed in quintal per acre in the study villages

Villages	Independent variable (X)	Dependent variable (Y)	b (slope)	a (intercept)	R^2
	Percentage of NIA to NSA	Agricultural productivity of Mustard seed in quintal per acre			
Saota			0.21	−8.56	0.99
Kurumba			0.21	−8.58	0.99
Goltikuri			0.16	−3.86	0.98
Binuria			0.21	−7.53	0.99
Ruppur			0.20	−7.34	0.99

Source Computed by the researcher

intercept (a) is −3.86 which means the regression coefficient shows variation below this constant figure.

In the case of **Binuria** village, the regression coefficient (b) is (0.21) where the equation is ($Y = -7.53 + 0.21x$). This shows that per unit increase of percentage of NIA to NSA will increase 0.21 unit of agricultural productivity of Mustard seed in quintal per acre in this village. The coefficient of determination (r^2) is 0.99, which means 99% of the total variation in Y (agricultural productivity of Mustard seed in quintal per acre) is being explained by X (percentage of NIA to NSA). The intercept (a) is −7.53 which means the regression coefficient shows variation below this constant figure.

In the case of **Ruppur** village, the regression coefficient (b) is (0.20) where the equation is ($Y = -7.34 + 0.20x$). This shows that per unit increase of percentage of NIA to NSA will increase 0.20 unit of agricultural productivity of Mustard seed in quintal per acre in this village. The coefficient of determination (r^2) is 0.99, which means 99% of the total variation in Y (agricultural productivity of Mustard seed in quintal per acre) is being explained by X (percentage of NIA to NSA). The intercept (a) is −7.34 which means the regression coefficient shows variation below this constant figure (Table 8).

8.5 Productivity of Til

Til (*S. orientale*) is the second most important oil seed produced in this area, and so it is largely cultivated in all the sample villages. The study shows that the productivity of Til is very low, ranging in between 3 and 4.5 quintals per acre, as compared to other crops. The farmers of Ruppur and Binuria village produce it with larger amount per acre. Moderate amount of productivity has been observed in Saota and Kurumba villages. The smaller productivity has been found on Goltikuri village. In most cases, the scarcity and inadequacy of irrigation, application of smaller amount of chemical fertilizer, and rare use of organic manure are the main reasons responsible for the lower levels of productivity of this crop.

8.5.1 Correlation Analysis Between Percentage of NIA to NSA and Agricultural Productivity of Til in Quintal Per Acre in the Study Villages

Positive correlation coefficient between the percentage of NIA to NSA and agricultural productivity of Til in quintal per acre has been observed in all the sample villages (Saota, Kurumba, Goltikuri, Binuria, and Ruppur), which means that both the variables are directly proportional to each other. The computed t values of all the sample villages (Saota, Kurumba, Goltikuri, Binuria, and Ruppur) are significant at 1 and 5% level of significance. Therefore, the null hypothesis is rejected and the alternative hypothesis is accepted indicating there is significant statistical correlationship between the percentage of NIA to NSA (X variable) and agricultural productivity of Til in quintal per acre (Y variable) (Table 9).

8.5.2 Regression Analysis Between Percentage of NIA to NSA and Agricultural Productivity of Til in Quintal Per Acre in the Study Villages

In the case of **Saota** village, the regression coefficient (b) is (0.10) where the equation is ($Y = -5.27 + 0.10x$). This shows that per unit increase of percentage of NIA to NSA will increase 0.10 unit of agricultural productivity of Til in quintal per acre in this village. The coefficient of determination (r^2) is 0.95, which means 95% of the total variation in Y (agricultural productivity of Til in quintal per acre) is being explained by X (percentage of NIA to NSA). The intercept (a) is −5.27 which means the regression coefficient shows variation below this constant figure.

In the case of **Kurumba** village, the regression coefficient (b) is (0.11) where the equation is ($Y = -5.56 + 0.11x$). This shows that per unit increase of percentage of NIA to NSA will increase 0.11 unit of agricultural productivity of Til in quintal per acre in this village. The coefficient of determination (r^2) is 0.95, which means 95% of the total variation in Y (agricultural productivity of Til in quintal per acre) is being explained by X (percentage of NIA to NSA). The intercept (a) is −5.56 which means the regression coefficient shows variation below this constant figure.

In the case of **Goltikuri** village, the regression coefficient (b) is (0.08) where the equation is ($Y = -2.56 + 0.08x$). This shows that per unit increase of percentage of NIA to NSA will increase 0.08 unit of agricultural productivity of Til in quintal per acre in this village. The coefficient of determination (r^2) is 0.88, which means 88% of the total variation in Y (agricultural productivity of Til in quintal per acre) is

Table 9 Correlation analysis between percentage of NIA to NSA and agricultural productivity of Til in quintal per acre in the study villages

Villages	Independent variable (*X*)	Dependent variable (*Y*)	*r* value	Computed *t* value	Tabulated value		Level of significance	Null hypothesis	Alternative hypothesis
	Percentage of NIA to NSA	Agricultural productivity of Til in quintal per acre			1%	5%			
Saota			0.97	15.28	3.06	2.18	1 and 5%	Rejected	Accepted
Kurumba			0.97	15.44	3.01	2.16	1 and 5%	Rejected	Accepted
Goltikuri			0.93	8.47	3.17	2.23	1 and 5%	Rejected	Accepted
Binuria			0.97	15.91	3.01	2.16	1 and 5%	Rejected	Accepted
Ruppur			0.97	17.24	2.98	2.15	1 and 5%	Rejected	Accepted

Source Computed by the researcher

Table 10 Regression analysis between percentage of NIA to NSA and agricultural productivity of Til in quintal per acre in the study villages

Villages	Independent variable (*X*)	Dependent variable (*Y*)	*b* (slope)	*a* (intercept)	R^2
	Percentage of NIA to NSA	Agricultural productivity of Til in quintal per acre			
Saota			0.10	−5.27	0.95
Kurumba			0.11	−5.56	0.95
Goltikuri			0.08	−2.56	0.88
Binuria			0.11	−4.80	0.95
Ruppur			0.10	−4.59	0.96

Source Computed by the researcher

being explained by *X* (percentage of NIA to NSA). The intercept (*a*) is −2.56 which means the regression coefficient shows variation below this constant figure.

In the case of **Binuria** village, the regression coefficient (*b*) is (0.11) where the equation is ($Y = -4.80 + 0.11x$). This shows that per unit increase of percentage of NIA to NSA will increase 0.11 unit of agricultural productivity of Til in quintal per acre in this village. The coefficient of determination (r^2) is 0.95, which means 95% of the total variation in *Y* (agricultural productivity of Til in quintal per acre) is being explained by *X* (percentage of NIA to NSA). The intercept (*a*) is −4.80 which means the regression coefficient shows variation below this constant figure.

In the case of **Ruppur** village, the regression coefficient (*b*) is (0.10) where the equation is ($Y = -4.59 + 0.10x$). This shows that per unit increase of percentage of NIA to NSA will increase 0.10 unit of agricultural productivity of Til in quintal per acre in this village. The coefficient of determination (r^2) is 0.96, which means 96% of the total variation in *Y* (agricultural productivity of Til in quintal per acre) is being explained by *X* (percentage of NIA to NSA). The intercept (*a*) is −4.59 which means the regression coefficient shows variation below this constant figure (Table 10).

8.6 Productivity of Mursuri

Mursuri (*E. hirsutum*) is the most important pulse crop produced in this area, and so it is largely cultivated in all the sample villages. The study shows that the productivity of Mursuri pulse crop is very low, ranging in between 4 and 5 quintals per acre, as compared to other crops in different study villages. Goltikuri, Kurumba, and Ruppur villages obtain better production from per acre of land by cultivating this crop. On the other hand, farmers of Saota and Binuria villages produce this crop with moderate amount of productivity. The study also shows that larger amount of productivity of this crop is obtained by those farmers who produced it with the help of regulated

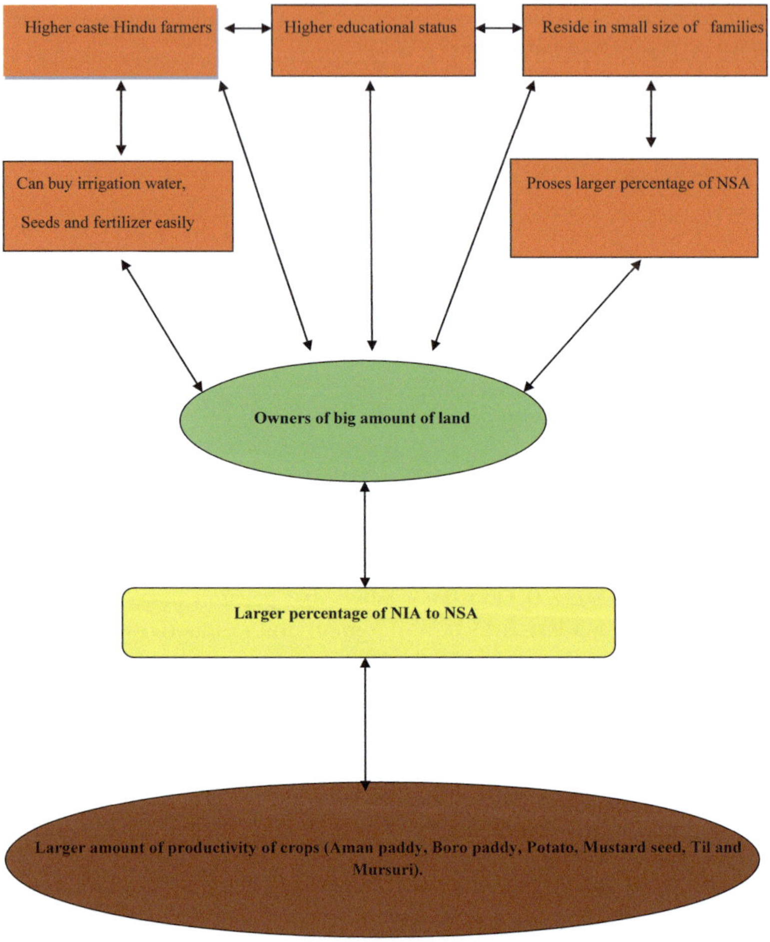

Fig. 2 Schematic Diagram showing the interrelationships among irrigation, agricultural productivity of crops, and different socioeconomic aspects of the farmers

supply of Submersible pump irrigation and Tank irrigation water and larger amount of application of NPK and Urea fertilizers (Fig. 2).

8.6.1 Correlation Analysis Between Percentage of NIA to NSA and Productivity of Mursuri in Quintal Per Acre in the Study Villages

Positive correlation coefficient between the percentage of NIA to NSA and agricultural productivity of Mursuri in quintal per acre has been observed in all the sample villages (Saota, Kurumba, Goltikuri, Binuria, and Ruppur), which means that both the variables are directly proportional to each other. The computed t values of all the sample villages (Saota, Kurumba, Goltikuri, Binuria, and Ruppur) are significant at 1% and 5% level of significance. Therefore, the null hypothesis is rejected and the alternative hypothesis is accepted indicating there is significant statistical correlationship between the percentage of NIA to NSA (X variable) and agricultural productivity of Mursuri in quintal per acre (Y variable) (Table 11).

8.6.2 Regression Analysis Between Percentage of NIA to NSA and Agricultural Productivity of Mursuri in Quintal Per Acre in the Study Villages

In the case of **Saota** village, the regression coefficient (b) is (0.07) where the equation is ($Y = -1.49 + 0.07x$). This shows that per unit increase of percentage of NIA to NSA will increase 0.07 unit of agricultural productivity of Mursuri in quintal per acre in this village. The coefficient of determination (r^2) is 0.95, which means 95% of the total variation in Y (agricultural productivity of Mursuri in quintal per acre) is being explained by X (percentage of NIA to NSA). The intercept (a) is -1.49 which means the regression coefficient shows variation below this constant figure.

In the case of **Kurumba** village, the regression coefficient (b) is (0.08) where the equation is ($Y = -1.34 + 0.08x$). This shows that per unit increase of percentage of NIA to NSA will increase 0.08 unit of agricultural productivity of Mursuri in quintal per acre in this village. The coefficient of determination (r^2) is 0.93, which means 93% of the total variation in Y (agricultural productivity of Mursuri in quintal per acre) is being explained by X (percentage of NIA to NSA). The intercept (a) is -1.34 which means the regression coefficient shows variation below this constant figure.

In the case of **Goltikuri** village, the regression coefficient (b) is (0.06) where the equation is ($Y = -0.22 + 0.06x$). This shows that per unit increase of percentage of NIA to NSA will increase 0.06 unit of agricultural productivity of Mursuri in quintal per acre in this village. The coefficient of determination (r^2) is 0.86, which means 86% of the total variation in Y (agricultural productivity of Mursuri in quintal per acre) is being explained by X (percentage of NIA to NSA). The intercept (a) is -0.22 which means the regression coefficient shows variation below this constant figure.

In the case of **Binuria** village, the regression coefficient (b) is (0.07) where the equation is ($Y = -1.14 + 0.07x$). This shows that per unit increase of percentage of NIA to NSA will increase 0.07 unit of agricultural productivity of Mursuri in quintal per acre in this village. The coefficient of determination (r^2) is 0.95, which means 95% of the total variation in Y (agricultural productivity of Mursuri in quintal per acre) is being explained by X (percentage of NIA to NSA). The intercept (a) is -1.14 which means the regression coefficient shows variation below this constant figure.

In the case of **Ruppur** village, the regression coefficient (b) is (0.07) where the equation is ($Y = -1.09 + 0.07x$). This shows that per unit increase of percentage of NIA to NSA will increase 0.07 unit of agricultural productivity of Mursuri in quintal per acre in this village. The coefficient of determination (r^2) is 0.95, which means 95% of the total variation in Y (agricultural productivity of Mursuri in quintal per

Table 11 Correlation analysis between percentage of NIA to NSA and productivity of Mursuri in quintal per acre in the study villages

Villages	Independent variable (X)	Dependent variable (Y)	r value	Computed t value	Tabulated value		Level of significance	Null hypothesis	Alternative hypothesis
					1%	5%			
	Percentage of NIA to NSA	Agricultural productivity of Mursuri in quintal per acre							
Saota			0.97	15.84	3.01	2.16	1 and 5%	Rejected	Accepted
Kurumba			0.96	12.48	3.11	2.20	1 and 5%	Rejected	Accepted
Goltikuri			0.93	9.80	2.95	2.13	1 and 5%	Rejected	Accepted
Binuria			0.97	15.33	3.06	2.18	1 and 5%	Rejected	Accepted
Ruppur			0.97	17.57	2.95	2.13	1 and 5%	Rejected	Accepted

Source Computed by the researcher

Table 12 Regression analysis between percentage of NIA to NSA and agricultural productivity of Mursuri in quintal per acre in the study villages

Villages	Independent variable (*X*)	Dependent variable (*Y*)	*b* (slope)	*a* (intercept)	R^2
	Percentage of NIA to NSA	Agricultural productivity of Mursuri in quintal per acre			
Saota			0.07	−1.49	0.95
Kurumba			0.07	−1.34	0.93
Goltikuri			0.06	−0.22	0.86
Binuria			0.07	−1.14	0.95
Ruppur			0.07	−1.09	0.95

Source Computed by the researcher

acre) is being explained by *X* (percentage of NIA to NSA). The intercept (*a*) is −1.09 which means the regression coefficient shows variation below this constant figure (Table 12).

9 Identification of Problems

Canal is an important source of irrigation. It plays a leading role in agricultural development, but Canal water is not available after kharif season for irrigation in the study area. Some part of canal has eroded (in the south side of Binuria, the northwest side of Kurumba, in the west side of Saota, and the south and west side of Ruppur).

During heavy monsoon rainfall, due to the uncontrolled flush irrigation, the water slopes down vigorously over the agricultural fields to cause soil erosion and water congestion. These problems not only degrade the land causing a decline in the productivity value of crop, but also increase the cost of production. The water also uproots large number of seedlings on a considerable part of the study area.

Tank irrigation is still an important source of irrigation in the study area. But these tanks are, in many cases, silted up, neglected and derelict, which have reduced considerably, the actual irrigated area. This problem affects adversely the crop productivity of the study area.

Due to regular hike in the price of diesel, the cost of operating shallow tube well and pumps for irrigation is becoming expensive day by day. The cost of irrigation is being increased for increasing the price of electric unit.

10 Policy Measures

Although the villages under study have adequate amount of water resources, the physical, socioeconomic, technological, and administrative problems hinder the management of these water resources, which consequently affects the economical status of the poor farmers to a large extent. So, the researcher has suggested some measures to overcome such problems.

Uneven lands must be leveled and adequate number of field channels should be constructed to distribute water systematically all over the command area with minimum wastage.

The release of water from the dams should be made in such a manner that maximum farmers receive benefits from them. Depending on the weather conditions, the release of water should be controlled. The number of canals should be decided depending on the capacity of the dams.

The price of diesel for agricultural use should be monitored and controlled by the government.

Farmers should construct underground reservoir with cover and preserve surplus rain water there for use during Rabi cultivation.

During Rabi season, the farmers can choose pulses crop to grow in lieu of Boro paddy to take the advantage of limited irrigation. This apart, Rhizobium Bactria is found at the roots of the pulses crop which increases the fertility of the soil.

Farmer should take training from the agricultural research center like pollisikha Kendra or Rotindranathkrishibigyan Kendra in sriniketan.

11 Major Findings

Aman paddy, Boro paddy, Potato, Mustard seed, Mursuri (*E. hirsutum*), and Til (*S. orientale*) crops are the gifts of irrigation in Bolpur Sub Division. Recently, the production of Boro cultivation has increased for the improvement of irrigation. During some years, people started to cultivate oil seed and Mursuri in place of Wheat. The study also reveals that the farmers who are the owners of big amount of land, posse's larger percentage of NSA, because such big farmers bring highest amount of their lands under the practice of cultivation due to irrigation facility. Those who have percentage of NIA to NSA less are most small and marginal farmers and they have more percentage of fallow land to NSA, because small and marginal farmers cannot buy Submersible water easily. Boro paddy is cultivated where Submersible pump is primarily available. Those who are big and medium farmers can buy Submersible water, seeds, and fertilizer easily and they produce more. Higher level application of chemical fertilizers and pesticide, timely and adequate supply of irrigation, and higher level supervision help to have an upper level productivity of Potato crop. The study also shows that larger amount of productivity of Mustard seed crop is obtained by those farmers who produced it with the help of regulated supply of Submersible pump irrigation and Tank irrigation water and larger amount of application of NPK, and Urea fertilizers.

12 Conclusion

Due to the advancement of science and technology, a major change in the irrigation potentialities is observed in respect of both major and minor irrigation schemes. As a result of that there is a major shift in land use, cropping pattern, and productivity of crops. As, for example, the amount of barren land has been reduced, whereas the same of NSA is increased considerably. Recently, the production of Boro cultivation has increased for the improvement of irrigation. More development of that area will be possible if a unity between agriculture and scientific irrigation infrastructure is created.

References

1. Das, N.G.: *Statistical Method (Part-I & II)*, pp 206, 297, 314. M. Das and Co., Calcutta
2. Government of India, Ministary of Agriculture and Irrigation, New Delhi: Report of the National Commission on Agriculture, Part-X, Inputs (1976)
3. Husain, M.: *Systematic Agricultural Geography*, pp. 186–255. Rawat Publication, Jaipur (2009)
4. Motebennur, S.S.: Impact of irrigation on agricultural development in Dharward district of Karnataka state. Int. J. Sci. Res. **2**(1), 119–121 (2013)
5. Todkari, G.U.: Impact of irrigation on agriculture productivity in Solapur district of Maharashtra state. Int. J. Agric. Sci. **4**(1), 165–167 (2012)

Big Data, Climate Smart Agriculture and India–Africa Relations: A Social Science Perspective

Ramnath Reghunadhan

Abstract Currently, climate change has emerged as the greatest challenge to humanity and pertains to possibility of events that have led to the previous mass extinctions on Earth. It is having a greater impact on agriculture through a multitude of challenges. In the emerging scenario of digital interconnectedness as well as complex interlinkages of climate change leading and increasing population stresses, there have arisen the need to find solutions. These solutions should not only deal with the pertinent issues, but also provide innovative and technological solutions as well. This is critical in the field of agriculture, on which populations of countries are dependent upon, both for production as well as for consumption. This paper analyses the possibilities of collaboration across agriculture and allied activities that can benefit both Indian and African economies by integrating big data into climate-smart agriculture, which thereby increases agricultural productivity as well as efficiency in the utilisation of resources. It takes exploratory and descriptive research, with the intention of filling the void of lack of literature in social sciences.

Keywords Big data · India · Africa · Climate-smart agriculture · Social science

1 Introduction

Currently, climate change has emerged as the greatest challenge to humanity and pertains to the possibility of events that have led to the previous mass extinctions on Earth. It is having a greater impact on agriculture through a multitude of challenges. In the current spatio-temporal paradigm the greatest challenge will be on feeding a growing population on that is dependent on limited resources, especially with degrading but constrained availability of land resources, transforming climatic conditions, diminishing rainfall as well as increasing inequality. The increasing pop-

Scopus Author ID: 57208542830.

R. Reghunadhan (✉)
Research Scholar, Department of Humanities and Social Sciences, Indian Institute of Technology Madras (IITM), Chennai, Tamil Nadu, India
e-mail: vrramnath@gmail.com

P. K. Pattnaik et al. (eds.), *IoT and Analytics for Agriculture*, Studies in Big Data 63,
https://doi.org/10.1007/978-981-13-9177-4_6

ulation, ambitious growth targets, increasing energy consumption levels and related as well as altering dietary requirements have but increased stress on natural resources with relation to the production of food and crop yields in a larger scale so as to avoid any actualisation of Malthusian prophecy.

However, with the increasing incidence of climate change, famine, hunger, exacerbating conflicts, resulting refugee influx to developing and even developed nations have altered the equation. With an estimated one-tenth of population still being undernourished, threatening the not just the food and nutrition security of citizens, but the social, economic, cultural as well as political stability of nations, which may often lead to the rise of authoritarian-illiberal democracies in not just developing, under-developed and least developed countries (LDCs) but in developed, traditionally liberal nations as well. This paper analyses the possibilities of collaboration across agriculture and allied activities that can benefit both Indian and African economies by integrating big data into climate-smart agriculture, which thereby increases agricultural productivity as well as efficiency in the utilisation of resources. This depends primarily on the capability building, adaptation as well as integration of innovative technologies into agriculture and other allied activities of farmers to climate-resilient but innovative technologies. In this context, the possibility of enhancing South-South cooperation through the partnership between India and African countries in the field of agriculture and allied sectors has been argued by analysing the possibility of imbibing the functionalities and uses of Big Data into the existing system of agricultural production as well as in institutionalizing it have been analysed. This paper analyses the possibilities of collaboration across agriculture and allied activities that can benefit both Indian and African economies by integrating big data into climate-smart agriculture, which thereby increases agricultural productivity as well as efficiency in the utilisation of resources. It takes exploratory and descriptive research, with the intention of filling the void of lack of literature in social sciences.

1.1 Emerging Scenario in Agriculture

Currently, in the global arena, in regard to the net farm output, India ranks second and is one of the largest exporters of food grains, especially to African nations. But similar to India's counterparts in Africa, India too has a high population pressure on land and other natural resources to meet its demands for food as well as development needs are concerned. The need to meet the food security, nutrition security as well as livelihood demands of the world's second largest populated country is a formidable challenge for any government or authority. This is due to the increasing challenges and issues of climate change, environmental degradation, salination, pollution, declining productivity in major cropping regions as well as a rapidly shrinking as well as stressed out resource base in the nation.

According to the report World Population Prospects: the 2017 Revision, the total population in the world is projected to reach 8.6 billion in 2030, 9.8 billion in 2050 and 11.2 billion in 2100, with India expected to become the most populated country by 2024. Among the 10 most populated countries, with increased fertility rate and poverty rate, African nations are expected to have the larger population, collectively closer to India and China [36]. This could not only result in an increasing stress on the existing infrastructure, exacerbate poverty in different parts, increase incidences of diseases, but coupled with the existing use of unscientific methods in urban planning and development activities, it will emerge as the greatest challenge across nations to achieve the Sustainable Development Goals (SDGs). The impact could probably create a ripple effect in multiple sectors as well as in different cross sections of population, especially in South Asian and African countries, mainly because of their lower agricultural productivity, increasing population and relatively lower infrastructural capacity to meet the demands. This will accentuate consumption rate due to the emergence of middle class as well as other development activities, with the population in developing, underdeveloped and least developed nations (LDCs) having to bear the greater share of the brunt.

Despite being the largest provider of jobs and livelihood to a large number of population in both India and Africa, agriculture is directly vulnerable to climate change as well as global warming. Any impact on agriculture and allied sectors can greatly affect not just the food or nutrition security of a nation, but can have a greater impact on the rates of unemployment as well as the stability of the nation or regime. It has a greater impact on the supply chains, not only locally or nationally but globally as well.

1.2 Scenario of African Countries

The agricultural ecosystems in Africa are mainly dependent upon rainfall as well as river system. The countries like Uganda, Burundi, Ethiopia and Burundi have major agricultural farms that are concentrated in highlands and are perennial in nature. This has a major influence type of crops (like wheat, barley), cropping patterns, plantation crops (particularly due to colonial influences) like bananas, coffee, root crops like cassava and sweet potatoes, beans as well as cereals are major crops. The agriculture and other allied activities are interlinked with the socio-economic as well as the biophysical characteristics and are mostly temperate form of the farming system. There is considerable mechanisation in major centres of farming but relatively archaic forms in other regions. The dependence on livestock is immeasurable, particularly when agriculture is affected due to climatic conditions. In the central and western regions of Africa, the treeroot system is followed by the population, while in eastern and southern regions of Africa the cereal-root crop system is being followed, something which exists for generations. These are prominent in countries like Ghana, Nigeria, Democratic Republic of Congo (DRC), Côte d'Ivoire, Burkina Faso, Benin, Sudan (Savannah) and in parts of Tanzania [12, 23].

Currently, African countries been hotspots of conflicts in the world, with multiple but deeply entrenched human-induced natural disasters being a common scenario. There are issues of food insecurity, poor quality in terms of nutritional value, adverse conditions for women and children, volatile prices of food and food riots and the like. In relation to the countries in the world, from both the West and the East, the African region has relatively the lowest productivity in agriculture and allied activities. An estimated two-thirds of the total population still depend on agriculture, among which a large number of the population are living below the one USD per day benchmark, with governments and regimes facing the biggest challenge in ensuring institutionalisation of a development policy that can provide inclusive but specific solutions to the local problems that can possibly, in any way, reduce if not end hunger, poverty and malnutrition in the region. Agriculture remains the important part of livelihood, economic and food and nutrition security of the rural poor in African countries and thus is a direct indicator in determining the standard of living of the population. The food production per capita in comparison to the international benchmark of optimum productivity has declined by in SSA. The productivity of cereals has largely remained stagnant, while the yield of other food crops and livestock has declined [13].

In the global market, a relatively low productivity of exports from Africa has greatly impacted the competitiveness, which has had a bigger impact on its balance of payment, and thus has increased the stress of debt on the public sector, thus affecting the social welfare programmes by the various governments and regimes. The various factors for this condition include the socio-economic intricacies, biophysical constraints, ineffectual policies as well as legislations adopted by various governments, unsustainable land management practices, sporadic rainfall, reduced access to usable water sources and diversion of fertile land for non-agricultural purposes all amounting to acute effects of climate change. This has actually resulted in relatively lower food productivity, have reduced the fertility of soil and stock of soil nutrients, lowered the quality of fodder, which have resulted in ensuing endemicity in livestock [10]. The increasing incidence of desertification, and the increasing profusion of arid and semi-arid regions and/or zones have but exacerbated the climatic conditions, particularly in Africa. The increasing temperatures has also drastically increased evapotranspiration, degraded the organic matter (and fertility) while reducing the waterholding capacity of humus resulting in deficiency of macro-nutrients in the soil [5, 8, pp. 669–680].

1.3 Scenario in India

In India, in terms of demographic, agriculture is the broadest economic activity that is intertwined collectively throughout the socio-economic fabric of the society. The country having witnessed phases of Green Revolution led to the successful production of surplus food grains in parts of the country as well as reduced the incidence of famine that was earlier prevalent during the colonial rule. The institutionalisation of Green Revolution thus changed this residual consequence in the post-independence period and reduced poverty and hunger among larger sections of the population. Thus, in the latter half of the twentieth century, India made impressive strides in agriculture, emerging from a status of country of famines to a powerhouse of foodgrain production [38].

A country with more than 1.3 billion population, being second in world farm output, sixth in agricultural exports, with largest cattle population, producing one-tenth of world fruit production India is among the leaders of agricultural production in the globe. The food production increased due to greater dependence on small landholding farmers, peasant families, coupled with inclusive policies and legislations, production and productivity strategies, investment in public infrastructure, research and development (R&D) as well as extension and outreach programmes for crops, livestock and fisheries have led to a significant increase in foodgrain production. The nature of the soil, climatic conditions, cultivation methods, techniques, irrigation, use of fertilisers, temperature conditions affable to different crop, plant and insect varieties, the prevalence of rainfall, harvesting and other related factors. Agriculture in India depends on the social, economic, cultural as well as climatic factors, many of which are unique (similar to African nations) [29].

The historically available crop yield information in India determined the operation of the supply chain in the country, particularly for industries that depended on agricultural products such as raw material, animal husbandry, poultry, seeds fertilisers and the like. Thus, it provides the largest livelihood security and rural employment opportunities to nearly half of the total population in India, particularly in the rural areas, but contributing only around 14% of the gross domestic product (GDP). However, it is also one of the emerging actors in sustainable agriculture, with a huge potential for organic farming, with particular focus on imbibing environmentally sustainable technologies [20]. This could actually reduce the degraded condition of lands used for agriculture, and in return ensue soil conservation, increase moisture retention capacities natural resource management (land and water) and thus create conditions for holistic rural development in the country.

The emergence of the middle class in India, with increasing income levels, has increased consumption levels and would also lead to a projected additional demand of 2.5 million tonnes of foodgrains on an yearly basis. The demand for foodgrains in the year 2020 is projected to reach around a level of 256 million tonnes, comprising of rice (nearly half of the production), wheat (one-third of production) and the remaining mostly accounts for the production of coarse grains and pulses. Besides, there will be a significant increase in demand for meat, fish, vegetables, fruits and other horticultural products (ICAR website). Incidently, agriculture and most importantly farming has emerged as an unprofitable industry over the last few decades. There has been increasing number of farmer suicides, and the state has found itself unable to unilaterally deal with the issue. According to an OECD report, taking 26 countries for the period of 2014–16, India was among the three countries that had negative farm revenues (a fall of 6%). The other two countries were Ukraine and Vietnam. Over a period of 2000–16, the country's gross farm receipts (or revenues) showed a decline of 14%. The major issue is the issue of distortion of prices as well as imbalance in supply and demand. The report found that more than three-fifths of the farming communities will be drawn into poverty if this trend continues [16]. A prominent reason for the agricultural crisis in India include the issues related to cropping pattern, soil quality, irrigation and monsoon, marketing, storage facilities, capital, use of fertilisers and biocides, quality of seeds and the like. Interestingly, with more half the population still dependent on agriculture and allied activities and issues of farmer suicides, poverty and undernourishment of a part of the population, India still needs to strive forward to work with technological advancements in dealing gaps and increasing the level of production, its supply, logistics, warehousing and balance the demand. This is possible through what is known as Big Data.

2 Emergence of Big Data

In the current global scenario, with the emergence of new, innovative and disruptive technologies including cyber-physical linkages, Internet of Things (IoT), all of which have pertinently resulted in an explosion of large data and information, commonly referred to as big data [31]. In the current spatio-temporal paradigm, by furthering research that primarily depends on upgrading collaborating and supporting joint R&D projects with countries could actually lead to what is known as the next 'Green Revolution'. Concomitantly, agriculture and the allied activities are undergoing the digital revolution, and even as small-scale farmers, crop growers and animal breeders are gathering information, if not passively but actively collected by using agricultural equipment. Many farmers will be able to use the information from large data sets and data analytics to create innovative but effective decision-making. The big data analytics techniques include advanced machine learning, advanced statistics, advanced data mining and the like [30].

A large number of global and transnational corporations have started introducing technological upheavals in agriculture, say the tractors with sensors that provide data about soil and crop conditions, even at real-time conditions to understand moisture content in the soil, climatic conditions with regard to the area or region, the amount of macro-nutrients or micro-nutrients that the soil needs, the demand for a crop in the market or future modelling of the yield that a land can provide [4]. All of this provide with a huge amount of data sets along with the instruments or tools that collect, aggregate, analyse and disseminate data, information and knowledge to all requisite stakeholders that can transform not just the interrelationships between the producer and consumer but the health security, nutrition security of population as well as the development of the nation as a whole. Within the next few years, the annual spending by the US-based companies on big data is estimated to be more than 100 billion USD, including investments in both products and services [24]. In two years, the big data market is estimated to more than double to 49.1 billion USD in 2020, become three times in 2021 and become four times in 2022. This shows an exponential growth annually for the next five years (Fig. 1). Figure 1 showing forecast of revenue of big data market around the globe between 2017 and 2022.

In the beginning of the twenty-first century, Doug Laney defined the term big data, which primarily encompassed 'data volume' (large size), 'data velocity' (high band-

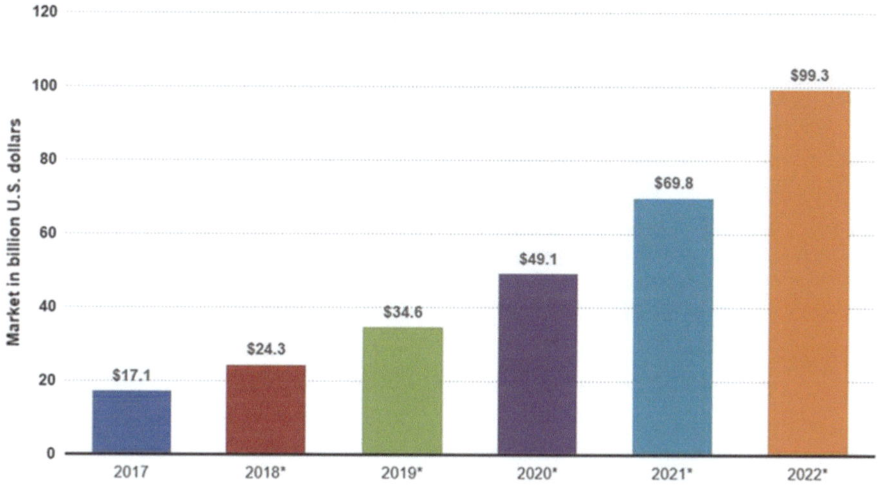

Fig. 1 Shows forecast revenue of big data market around the globe between 2017–22. *Source* Columbus [6]

width) and 'data variety' (data types and semantics) [22]. It designates data sets as usually characterised by their large size and higher complexity, thereby challenging the methodology used for traditional data analysis. It refers to a large but complex collection of information (or datasets) that needs to be processed by instruments or tools in ensuring accessible knowledge. It also comprises data (enterprise, social and/or machine) from the past or at real time and thus helps produce new knowledge, which can supplement, complement or even displace the existing knowledge or practices of different sectors. All aspects of data management, i.e. from data acquisition, capture, storage, transfer, access, sharing, modelling, learning, visualisation or human interactions, interpretation and finally the capacity to support reliable and accurate decisions can be undertaken by using big data. According to reports, the big data analytics sector in India is expected to increase by around 14 billion USD by 2025 and will register a growth rate of 26% over the next five years. India is currently among the top ten data markets in the world and could rise to top three in 2020 [21]. The applications for big data analytics include social analytics, mobile analytics, living analytics, video and visual analytics, text analysis and audio analysis [30]. Figure 2 shows the estimated demand for big data in various Indian cities.

The world is now more interconnected and creating a massive amount of data, which can be used for relatively more effective decision-making for the agricultural producers. It has transformed agricultural businesses and enterprises, with each point of critical information leading to more efficient use of resources. Based on the survey by Venturesity, a Bangalore-based online university, the estimated demand for skilled labour in big data in India is on the rise. Among the cities, Bangalore leads with 38% followed by Hyderabad at 14%, Pune at 11%, Gurgaon at 10%, both Mumbai and Chennai at 8% and finally cities Ahmedabad, Delhi, Noida and Vizag ranging

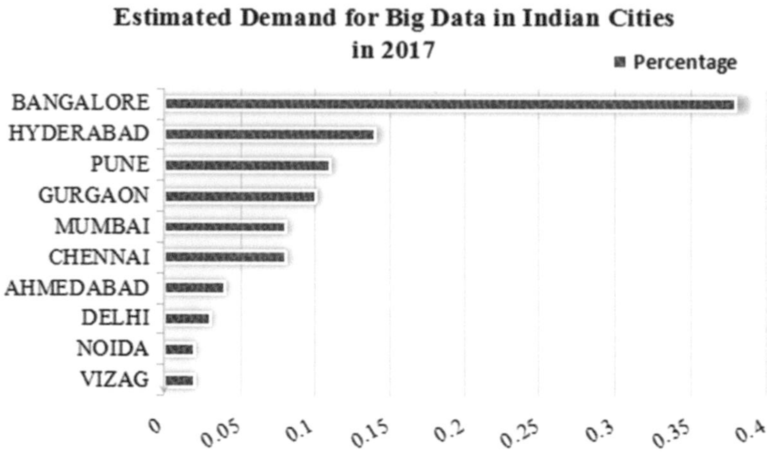

Fig. 2 Estimated demand for big data in Indian cities. *Source* Steinhart [33]

from 2 to 4%. The salary of big data analytics is relatively huge in relation to jobs in other sectors. A large number of technology giants like Google, Facebook and Amazon are also recruiting from India [33]. As technology is disrupting diverse areas of the existing agricultural supply chain, it also helps create opportunities to feed an increasing population within the country. Since the world is now more interconnected and creating a massive amount of data in much lesser time, big data can be used for relatively more effective decision-making among the agricultural producers. It transforms agricultural businesses and enterprises, with each point of critical information leading to more efficient use of resources.

A number of software and data analytics tools are needed in carrying out big data analysis. The ones that are built on open source principles, i.e. the ones that are essentially in the public domain and are free to use and for any purpose. *Hadoop* is one example for this. It is a collection (or framework) of analytics, software tools, technologies and/or applications designed for allowing organisations (of any number, size, structure) to collect, store and analyse huge amounts of data and information. Thus, it is designed to run at a relatively low-cost (ordinarily available) hardware. This reduces the need for hardware or equipment of exorbitant price or that have to be specially designed for a particular purpose only. Besides, there is *Map-Reduce*, *NoSQL* databases, graph databases and analytical frameworks. As technology is disrupting diverse areas of the existing agricultural supply chain, it also helps create opportunities to feed an increasing population not only within the country, but outside it as well. This is particularly important in reducing the intensity of poverty in the underdeveloped regions of India and African countries.

2.1 *Integrating Big Data into Climate-Smart Agriculture (CSA)*

Currently, CSA encompasses an approach that ensues steps in a direction towards achieving the objectives of food and nutrition security among the population, especially in the backward regions. It ascertains the aim of both environmental sustainability as well as socio-economic development. In India, with more than six lakh villages and nearly 130 million farmers in 127 agro-climatic zones, speaking in diverse languages, the challenges to bring out a cohesive universal language of farming among these pluralistic societies, is a great challenge. But integration of big data into CSA could provide a one-window solution for increasing the opportunities increasing the yield and productivity of more than 140 million hectares (ha) of agricultural land and using more than three thousands of crops the country [2]. The emergence of strategies and actions to imbibe CSA into agriculture has been a game-changer. According to the Food and Agriculture Organization of the United Nations (FAO):

> Climate-smart agriculture (CSA) is an approach that… [helps] transform and reorient agricultural systems to effectively support development [as well as] ensure food security in a changing climate. CSA aims to tackle three main objectives: sustainably increasing agricultural productivity and incomes; adapting and building resilience to climate change; and reducing and/or removing greenhouse gas emissions, where possible [11].

It also helps increase the productivity of agricultural lands, increase the resilience of crops and plants to drought, pest attacks, enable societal and economic development and integration and ensure livelihood security, food security and nutrition security to different stakeholders. It also helps achieve the directives of Sustainable Development Goals (SDGs) like reduction of poverty, zero hunger, good health, well-being, decent work, increased gender equality, economic growth, sustainability in living of communities, reduced inequalities, climate action, peace, justice, strong institutions in the coming decade, while focusing on environmental sustainability, with effective landscape management systems, while helping farmers adapt to productive climate-resilient and low-emission agricultural technologies and practices for agricultural systems, so as to ensure sustainable increases in food production. But it also seems to have loopholes in effective implementation, in achieving the dual aim of agricultural development as well as in dealing with climate change in a sustainable manner.

Now, as technological revolutions are taking place at a rapid pace, the emergence of big data analytics is the focal point to foster a Digital Green Revolution in agriculture. The vital importance of big data is that it has evolved as a form of technology that solves problems on the basis of historical data, data that are machine-generated as well as streaming of real-time data. It has generated a large quantity of data that are relevant to various stakeholders. This has been particularly possible by the use of sensors in and around the agricultural ecosystems. The digitalisation agriculture will provide real-time decision-making not just possible but efficient and effective as well. Figure 3 showing the uses of big data analytics for CSA is given below.

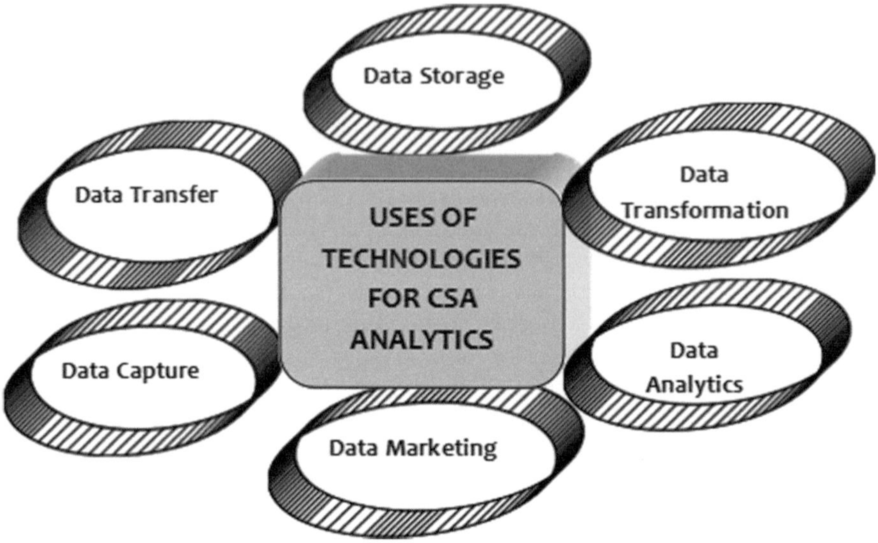

Fig. 3 Uses of big data analytics for climate-smart agriculture analytic. *Source* Gupta [15], Wolfert et al. [37, pp. 69–80]

Figure 3 shows various uses of big data analytics in CSA. Concomitantly, the uses include *data capture*, *data storage*, *data transfer*, *data transformation*, *data analytics* and *data marketing*. In *data capture*, the emphasis is on capturing data from digitised farms as well as warehouses with sensors, open data, biometric sensing, reciprocal data and genotype information. There are some key issues in this, namely the positioning of sensors, availability, quality and formats that are captured. In *data storage*, the emphasis is on cloud-based solutions as well as platforms, hybrid storage solutions, Hadoop (Hadoop Distributed File Systems [HDFS]), hybrid storage systems, cloud-based data warehouses and also what is known as data lakes (centralised repository with structured as well as junk data). But there are issues like easy accessibility, security, encryption vulnerabilities and cost of maintaining the infrastructure. In *data transfer*, the emphasis is on cloud-based platforms, wireless connectivity options as well as linked open data. There are inherent issues of safety, consent, responsibilities as well as liabilities, particularly on privacy and sensitive data. In *data transformation*, the emphasis is on normalisation, visualisation and anonymity in terms of processed data, machine learning algorithms. There are issues in heterogeneity of sources of data, automation in regard to cleansing or cleansed data as well as for preparation. In *data analytics*, the emphasis is on cognitive computing, yield models, ontology-based decision support systems (DSS) and standardisation as well as benchmarking solutions. There are key issues like semantic heterogeneity, in real-time analytics as well as scalability on the basis of the amount of work handled. And finally in *data marketing*, the emphasis is on data visualisation but there are inherent issues like privacy, ownership and new business models [15, 37, pp. 69–80]. Figure 4 shows flowchart of intelligent processing of agricultural big data.

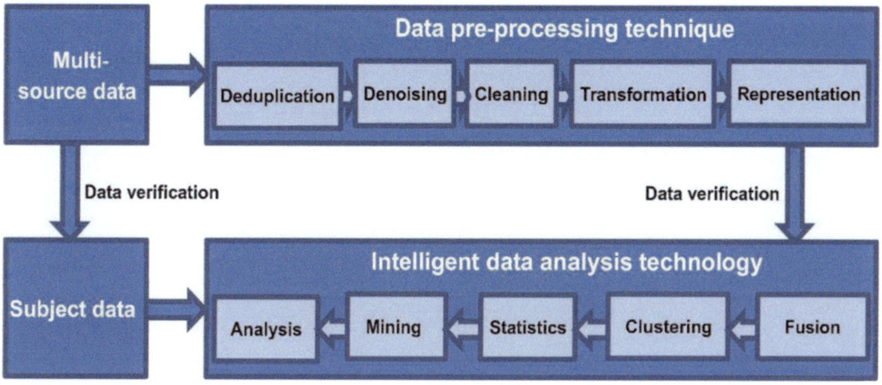

Fig. 4 Shows flowchart of intelligent processing of agricultural big data. *Source* Wolfert et al. [37]

Currently, big data has the potential to add value at each level that begins from the identification and selection of agricultural inputs, understanding variable parameters like soil and land elevation that is possible with the use of related future modelling, monitoring the moisture content in soil and air, yielding monitors, demand tracking, simulation, projection and extrapolation of possible price fluctuations, streamlining irrigation processes and the like. It also encompasses what is known as precision agriculture supplemented by extension activities of using geographic information systems–global positioning system–remote sensing technologies (GIS-GPS-RS) linked to mobile apps, profiling and sharing of information in social media to reach cultivators, farmers and different stakeholders as well. Figure 5 shows the flow of data and how integrating components of big data and climate-smart agriculture works.

This can help farmers increase their yields, decrease costs as well as save time. This also leaves a possibility of further bridging the digital divide, not only within the country but also between countries to enable sustainable South–South cooperation. It can also help in inventorying traditional as well as indigenous knowledge, best practices particularly suited for a location or region and provide information sharing between different stakeholders. This helps refine as well as redefine farming techniques and often ensues an increased interest in agriculture, possibly looking at it as a form of entrepreneurship creating possibly the 'Next Organic Green Revolution'. Figure 6 below shows an estimate of the amount of data generated by an average farm in a day.

Figure 6 shows an estimate of the amount of data that could be generated by an average farm per day. Within the next few years, the challenge from big data enhances with data points doubling every two to four years. The increasing number of data points means an increasing number of indicators to reach the final conclusion, which by the way leads to much efficient decision-making system. It also brings effective forecasting, input of accurate information and optimisation in use as well as effective protection of natural resources. Big data can transform agriculture in multiple ways. The use of sensors in fields, taking granular data points on soil con-

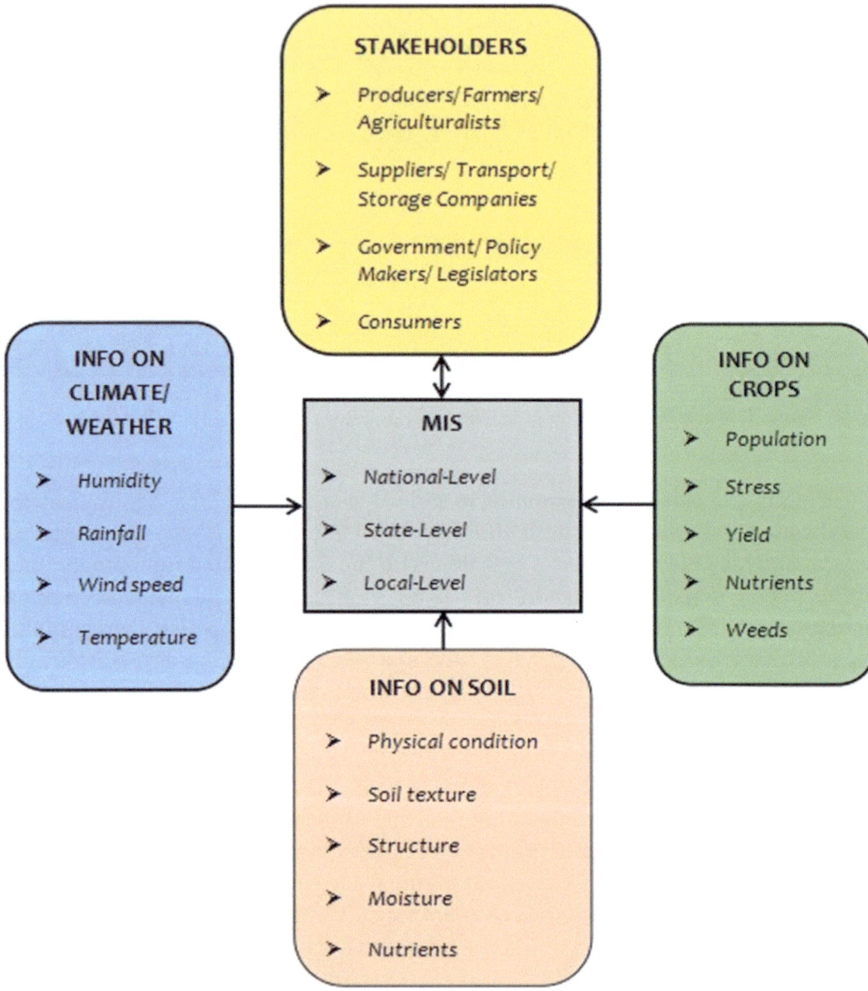

Fig. 5 Integrating components big data and CSA. *Source* Compiled by the Author; Note: *MIS* management information system (database)

ditions, wind, fertiliser requirements and water level as well as pest infestations. The use of seamless integration of data or information for stakeholders ensures district- and local-level planning will ensure more effectiveness in implementation. This is possible by making optimum predictions of seasonal and long-term variations climate, thereby increasing instances of effective predictability of crop yields. It will enable the creation of inclusive agricultural policy decisions, with detailed, subjective assessment of drought(s), extrapolate possible famines, availability of food grains and possible contingency planning that is required. They could further give detailed mapping of agricultural land use, production statistics and precision agricul-

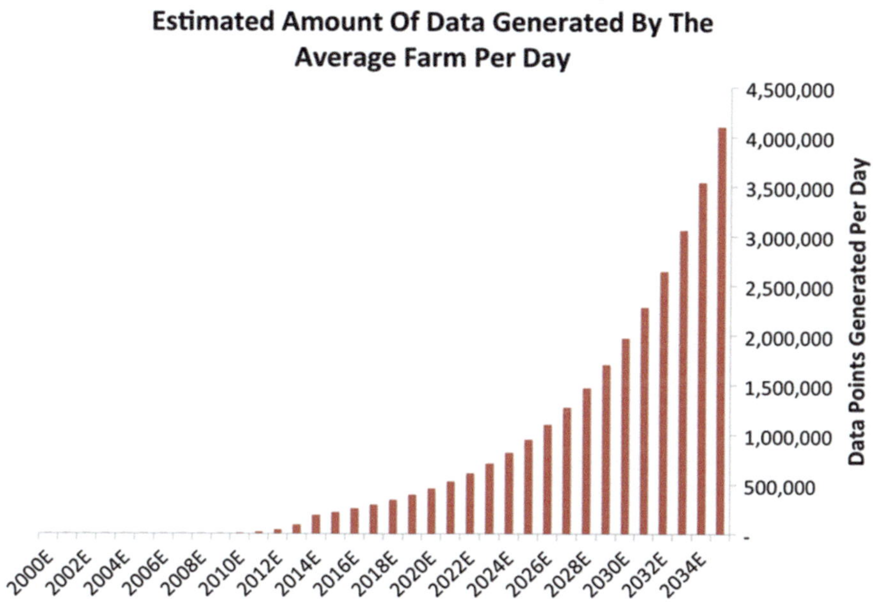

Fig. 6 Estimated amount of data generated by the average farm per day. *Source* Meola [25]

ture and help schedule demand-based irrigation, support cropping system analysis, provide damage assessment, pest or disease surveillance and help crop inventorying processes. Further information on land degradation helps in soil reclamation, conservation, for land use planning, increase carbon sequestration reduce greenhouse effect and the like. This helps provide food and livelihood security for farmers and fishing community, reduce over-exploitation of common property resources, help reduce soil erosion, improve soil fertility, soil restoration techniques and technologies, conservation tillage and residue management and improve farming or cropping system and agricultural intensification. Figure 7 shows a real-time nitrogen field advice through a cloud service.

The use of global positioning system (GPS) units on tractors and agriculture-related vehicles, tools and instruments can help optimise their usage. The use of data analytics can help improve decision-making, unmanned aerial vehicles (UAVs), drones and/or nanobots can patrol crop lands, collect data aerially and alert the farmers with regard to condition of crop or regarding possible problems, even by using future modelling. The RFID-based systems can trace as well as provide real-time stream of data stream with regard to farm products up to the end-user of the supply chain. The emergence of user-defined but location-specific software updates in CSA could help greatly improve yield, help in field mapping, crop scouting and weather forecasting, with information on deficiency of nutrient or water being sent to an app on the smartphone of the stakeholder. Big data can be used to analyse a wide variety of seeds across numerous fields, soil types and climates, help farms identify and analyse

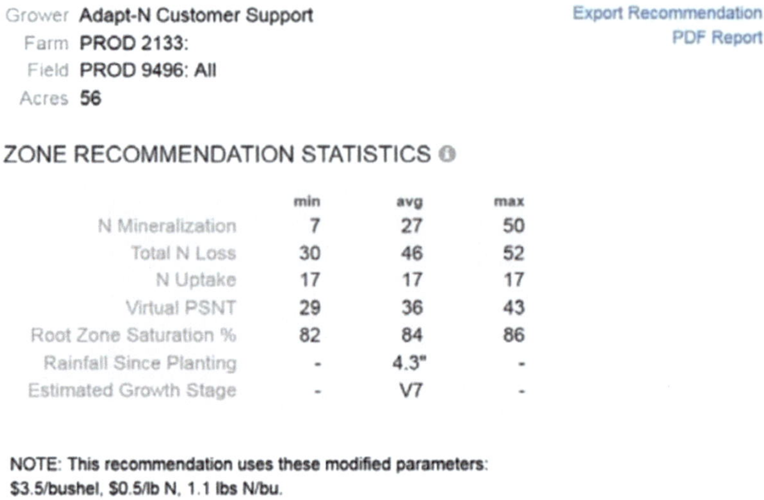

Grower Adapt-N Customer Support
Farm PROD 2133:
Field PROD 9496: All
Acres 56

Export Recommendation
PDF Report

ZONE RECOMMENDATION STATISTICS

	min	avg	max
N Mineralization	7	27	50
Total N Loss	30	46	52
N Uptake	17	17	17
Virtual PSNT	29	36	43
Root Zone Saturation %	82	84	86
Rainfall Since Planting	-	4.3"	-
Estimated Growth Stage	-	V7	-

NOTE: This recommendation uses these modified parameters:
$3.5/bushel, $0.5/lb N, 1.1 lbs N/bu.

Fig. 7 Real-time nitrogen field advice through a cloud service. *Source* Vanes and Woodard

crop-related diseases during the pre-harvest period. When integrated into CSA, it can aid farmers in ensuring highly effective water management system, reduce wastage, increase production, improve economic wherewithal of farmers and peasants as well as reduce the impact on the environment. It also brings effective forecasting, input of accurate information and optimisation in use as well as effective protection of natural resources. The data, in the form of maps of resources, land, water resources, delineated boundaries, differential layering of soil, help in further planning, for effective implementation as well as infrastructure development in agriculture in each area or region. It thus encompasses the socio-economic data and helps in effective climate governance even at the grassroot level, with an emphasis on bottom-up planning, by even increasing the standard of living among the common as well as indigenous population. This ensures a participatory development of the population at multiple levels and even enables reclamation, use and monitoring of agricultural land in a more productive way.

The well-being of indigenous and local population as well as their knowledge can be enhanced, often supplemented by welfare activities by both the government as well as NGOs. The particular need for expanding policy capacities as well as in broadening policy horizons can be ensued by providing overlapping solutions and benefits to communities, bringing out unexpected policy dimension(s) as well as reducing or countering scarcity-induced conflict in different areas. It helps create and evaluate information, augment the creation of a legal framework that is fair, which is enforced impartially, through transparent and accountable mechanisms. The built-in transparency will enable free flow of data, information as well as knowledge, all of which is disseminated to the end-user individuals and/or communities. The institutions, processes and data are accessible directly to all stakeholders, with clear mechanisms to understand and monitor them. This will enable greater gender-neutral participation in the decision-making mechanism(s) and process(es), thereby increasing the opportunities to better or maintain their well-being, augment broad-based participation, constructive capacity building and training as well as awareness.

3 Emerging Challenges

A huge challenge in regard to investment in Africa would be from China. The Chinese investment in Africa is behemoth and very huge for India to compete with. But interestingly, the three countries have been subjected to colonial subjugation in the past and thus represent a post-colonial shift in development and capital production. As for disrupting Chinese 'imperialistic' designs, an increasing level of cooperation between India and Africa at bilateral, regional and multilateral level seems could potentially diversify options for Africa and India as well. But this has often faced challenges due to internal roadblocks from administration, political bickering that hinder economic activities, lack of economic surpremacy comparable to China and increased importance of Indian state towards other actors. There are also roadblocks in developing CSA practices that are tailored to the needs and location-specific characteristics of farming systems, particularly in regard to the socio-economic conditions of population, the requirements of farmers, the agro-ecological context and the climatic conditions.

The need for providing immediate and long-term benefits from CSA practices in a sustainable way has been hindered, as the adoption of CSA practices is currently determined by the effectiveness of training provided, through seminars, workshops, sessions as well as through farmer-to-farmer learning (mouth-to-mouth knowledge) or documented knowledge that is either too slow to be disseminated in a larger scale, whose effectiveness can only be known through trial and error method by different stakeholders. This often leads to trust deficit among farmers and other producers, who may not readily agree to forego their methods and processes to possibly effective ones. This has also greater impact on the planning, designing and implementation of CSA programmes, extension activities, adopted strategies, for ensuring availability of credit and other inputs, infrastructure development as well as in investment plans. The gap between policy-makers and the local community and/or bodies can increase issues of trust deficit (mostly mutual) that can lead to differences as well as conflicts in the utilisation of resources, increasing leakages on inputs and benefits to farmers and producers, increasing ineffectiveness of subsidies, and in extreme cases altering the agricultural landscape of an area or region.

In the world of increasing concern for climate change, poverty incidences and conflict for resources, big data holds the key providing a viable, long-term and timely solution to the human race. It can provide access as well as gain access from farmers, producers, breeders, policy-makers and consumers as well as different related stakeholders. But there exists challenges to this new phenomenon, particularly in the initial times where the stage is being set. The collection, aggregation, management as well as analysis of data are becoming increasingly complex and time-consuming, which often hinders the activities to effectively produce knowledge as well as its dissemination. The assimilation as well as incorporation of these data, information and knowledge requires large but complex algorithms and data analytics software systems that are often developed and designed by agcompanies and transnational corporations, thus possibly institutionalising inherent discrimination in terms of accessibility to individuals, groups or communities and even nations that would lead to dictation terms by a few selected elite companies and/or nations [14].

In relation to digitalised agri-ecosystems and data systems, there are issues of data mobility; partnerships at individual, company, governmental and intergovernmental level; information sharing; adaptability; lack of requisite infrastructure; data ownership, security and sovereignty; as well the lack of trust among different stakeholders, particularly the farming communities. There are issues of limited of open data availability (open-source data) that hinders public interest, particularly that of a farmer at the grassroot level. The increasing incidences of data breach and compromising of security has increased concerns on the possibility of protecting information. It has also raised privacy concerns in the context of a lack of policy framework in order to regulate collection, management and usage of big data. The existing lack of infrastructure, lower penetration of Internet and broadband as well as increasing hesitation from farmers and indigenous communities to use as well as provide data and information are greater bottlenecks that are needed to be handled. The absence of a legal framework or institutionalisation of standardised procedures and security practices in protecting sensitive personal data or information (SPDI) within India (discussion have started on a draft Data Protection Bill) and African countries (with the exception of South Africa), and between them (unlike with Israel) can act as a major hindering issue in furthering cooperation in the field of big data [32].

But a large number of smallholding farmers or peasants, most of them usually unskilled for other jobs, ill-equipped with dealing with or availing knowledge and/or skill, coupled with the increasing complexity of ecological issues, degradation of available land and water resource, increasing climate variability have but exacerbated the risk of a large number of families falling into abject poverty. Another pertinent problem is the miniscule number of skilled scientists or researchers in different nations, who know or are capable of creating as well as executing the related algorithms capable of analysing large of amounts of data sets. Big data, like all data, has to undergo scrutinisation for ensuring content quality before it is being used in a large scale, so as to avoid not just ineffectiveness in its implementation but any possibility of catastrophic outcomes. There exists a bias and/or imbalance in the scale, accuracy and precision of data coming from different algorithms or filters that are but mostly created based on the perception of the inventors on what is to be analysed or neglected as well as the interpretation of the data that comes through. Thus, the subjectivity of analysis will be based on the perspective of what one considers to be normative (hermeneutically speaking) would not always provide the intended outcome. The adoption in developing and underdeveloped nations, specifically in the case of India and African would be slow and difficult in relation to developed nations, with lower population as well as increasing implementation of capital-intensive technologies.

4 Big Data-Integrated CSA and the Future of India–Africa Relations

There will be a huge strategic realigning around the globe and within India and Africa as well. A huge possibility that unveils with this South–South cooperation is the increase of African exports to USA, EU, China, South America as well as Asia. Interestingly, the Pan-African e-Network Project (PAENP), a brain-child project of the former President of India, the late Dr. Abdul Kalam (in September 2004) was formally launched in 2009, and is 'still' the backbone of Internet infrastructure in the African continent. The PAENP makes it viable for integrating big data and CSA in Africa and can be revamped through a joint initiative, particularly in the 47 learning centres (LCs) involving the five Regional University Centers in Ghana, Uganda, Cameroon, Egypt and Malawi. Besides cooperation in trade, technology transfer and investment promotion, Africa has provided its wholehearted support for India's permanent membership in the United Nations Security Council (UNSC). Currently, India and Africa have constructive discussions on 'food and energy security, climate change and trade'. India has offered assistance in entailing an '[African] Green Revolution through holistic capacity building in agricultural production', supply, 'storage and transportation"' as well as 'strengthen human development'. India is funding the projects under the Indian Technical and Economic Cooperation (ITEC), Know India Programme and Special Commonwealth Assistance to Africa Programme (SCAAP) [19, 26, 27].

This foreign policy decision by India seems to have a positive sum game for both the actors. For India, the India–Africa cooperation in investment and technology transfer in big data could have positive impact in the sense of low-cost agricultural investments and high returns for Indian entrepreneurs, increased imports from Africa providing for greater assurance of food and nutrition security to India, increased forex returns from Indian companies and individuals from Africa, enhances Indian stature in African continent as well as around the globe and provides for furthering Asia–Africa Growth Corridor. For Africa, the cooperation will increase agricultural productivity, labour and employment opportunities in various regions, increase standard of living, investment capital, currency value, development, education, skilling, energy security (alternative fuels) as well as reduce poverty and corruption among African population. This could also increase political stability, governance, grassroot-level innovation, emergence of small and medium enterprises (SMEs) and socio-economic security, food security as well as nutrition security. This is critical in various African countries that are hit by famine, food insecurity and famine-like conditions.

An increasing number of agfirms are in the process of adopting and integrating big data technologies, to gain insights from the massive quantity of diverse data. This could help understand, resolve and manage problems in a real-time basis and enable plugging gaps in knowledge that is often lacking in non-indigenous farming communities, groups, organisations and companies. In India, the private sector companies in the field of agribusiness, agri-start-ups and agri-supply chains are looking at the possibilities of using big data in agriculture. This has become prominent even among conventional business due the emergence of the climate change narrative around the globe. The use of farming data is a powerful tool, particularly in the hands of corporates as well as different agricultural companies, groups and start-ups. The lack of farmer representations in this level has left a gap in the existing race to claim authority over cropping or farming data that can lead to future ramifications for the farming populations. Currently, large companies like ITC, Tata Group, Wipro Group, Mahindra & Mahindra, McCain are developing solutions in transforming CSA in the country. Besides, in 2017, agstart-ups based in India have begun seed funding and gaining mass capital mobilisation due to the demand in the country. The start-ups include AgroStar (Pune), Farm Taaza (Bengaluru), CroFarm (Bengaluru), Aarav Unmanned Systems (Bengaluru), Aibono (Chennai), CropIn (Bengaluru), Gold Farm (Bengaluru), Ninjacart (Bengaluru), Waycool (Chennai), EM3 Agri Services (Noida), Intello Labs (Bengaluru), FarmLink (Bengaluru), Gramco Infratech Pvt. Ltd. (Indore) and Tessol (Mumbai). The issues of data accessibility, data ownership as well as data sovereignty are being hotly debated amongst different stakeholders in the international and national forums, but without much ado. The Supreme Court of India as well as other policy decisions, regulations and rules imbibe this in the latest of events. Interestingly, there has also been an emerging consensus that these technological development and use of data should be regulated by the government and enable the integration into policies that are to come by. The issues of data accessibility, disinformation and misrepresentation should be dealt with [3, 17, 28].

Besides the Indian private sector in Africa seems to have provided great inroads into Africa, unlike the Chinese investor Indian investments seems to focus on providing employment opportunities to African labourers and have reduced debt burden for African countries. The recent Chinese Belt and Road Initiative (BRI) seems to create debt trap for Africa which seems to increase political instability, societal tensions, economic drain and even intrude (directly and/or indirectly) into the foreign policy decision-making (processes) of various states in Africa. India and Africa need to increase investment in agriculture, imbibe and develop newer technologies for innovation and modernisation in agriculture as well as increase inclusiveness for small-holder agricultural families. The Delhi Declaration (2008) was the first Track 1 initiative by the Indian Government for exploring the possibilities and potential for increasing food production in Africa, particularly in sub-Saharan countries where climate-induced desertification and poverty is rampant among large behemoth of populations. The common characteristics of agricultural ecosystems in both India and African actually create a large number of opportunities for closer collaboration. Currently, African has a huge percentage of the cultivable land in the world, but is also able to generate one-tenth of global agricultural output, which is meagre in relation to demand. The possibilities of African countries to become some of the fastest growing economies could enable Indian agriculture and food sector to tap into the potential, thereby increasing by manifold the exports from African nations, and in providing labour-intensive alternative to the 'West'. An effective system is needed to deal with climate change that which introduces agriculture as a platform in dealing with climate-resilient activities, which can help mitigate issues as well as bridge gaps relating to agricultural resilience, food and nutrition security, provide a sustainable path for agricultural development and assimilate steps and processes to deal with climate change for all stakeholders.

In the current scenario, with increasing population and a slowing rate of growth (attuned in the global level), India needs to streamline the use of resources, induce domestic demand, improve accessibility of both the producers as well as consumers, circumvent the unwanted interventions by intermediaries, reduce leakages in income distribution and tap into the purchasing power of the population for overall development of the country. This is possible by investing on agricultural research facilities and training programmes in Africa, imbibing into the universities and curricula the certificate, diploma or masters courses related to big data, data analytics, data science, machine learning and artificial intelligence that focuses on professions like data analyst, data scientist, data architect, business analyst, database administrator, data management and governance (SMEs focused also). There should be an increasing information sharing and collaboration for CSA for both nations and in enabling streamlined value chains and partnerships for capacity building and technology transfer in the agricultural sector. There should be an enabling interface for governments, industry and academia (particularly track diplomacy) to increase the foreign direct investments (FDIs) and increase concessional lines of credit from Government of India (GoI). It will enhance the commitment to an open data strategy, making data available for scientific purposes and for public use, while supplementing conditions for the creation of an efficient information and innovation ecosystem that keeps pace with the challenges of our world. In the current scenario, the prominent issues of identity, data storage and accessibility need to be brought under well-defined standards, both legal and procedural, which is possible only when India accelerates the development of a data security policy.

In the case of private companies or corporates handling the data, the government has to take proactive measures and regulations in protecting the individual's identity as well as information. The government should also bring-forth stringent jurisdictional rules and monitoring mechanism, as well as penalise for any violation or misuse of data. Under the label of 'nudging' or 'Big nudging', many corporates are trying to influence citizens or netizens through a relatively modern form of paternalism. This issue becomes prominent with the arrival of virtual spaces which is transboundary in nature. The policy clearly needs to address standards around identity and access management, most pertinently on data storage, data sovereignty and encryption [18]. Globally, the major players in the field of big data in CSA include John Deere, AgDNA, Monsanto, BLUERIVER Technology, aWhere, Agribotix, OnaFarm, FarmersEdge, The Climate Corporation and the like.

India could enable the transfer of technology and capabilities from the 'haves', and in return tap into their technological prowess, which is of high priority with the emergence of accelerated trends of digitisation. But the right of an individual regarding personal information in big data should generally be in agreement with the Data Protection Directive of the European Union (1995) and not violated as far as possible. Recently, the EU has also brought out the General Data Protection Rules (GDPR) that has huge implications on companies with activities in EU countries. The huge quantity of information and data sets are available mostly in the form of what is called unstructured data (junk data) that have to be streamlined. It can advance the agriculture and extension services as well as create an inclusive ecosystem encompassing the needs and aspirations of the population, increase efficient participatory subsistence system, that which incorporates 'smart data warehousing' for effective decision-making as well as implementation. India has a strong information technology (IT) infrastructural base with relatively lower wage cost than developed nations, which gives a comparative advantage in exporting big data analytics services. India also possesses one of the world's largest number of IT-skilled professionals in the world, and thus the capability to imbibe big data analytics into food production, soil fertility as well as in dealing with climate change. The government should undertake steps to ensure the protection of basic rights of citizens, within an appropriate regulatory framework, code of conduct that for data algorithms and guarantee informational self-determination for the creation of a participatory digital society [1]. The creation of an enabling interface for governments, industry and academia increases the foreign direct investments (FDIs) and increases concessional lines of credit from Government of India (GoI). It will enhance the commitment to an open data strategy, making data available for scientific purposes and for public use, while supplementing conditions for the creation of an efficient information and innovation ecosystem that keeps pace with the challenges of our world.

The emergence of user-defined but location-specific software updates in CSA could help greatly improve yield, help in field mapping, crop scouting and weather forecasting, with information on deficiency of nutrient or water being sent to an app on the smartphone of the stakeholder. The use of telematics technology, provision of work instructions to machines, use of cloud data, availability of seed choices in practices, benchmarking, information sharing, livestock information at real-time, effective analyses, streamlining and dissemination of volumetric data including retail data, better possibility for gene editing, decision-support system and better genetic and genomics in management decisions. Big data can be used in analysing a wide variety of seeds across numerous fields, soil types and climates, helping farms identify and analyse crop-related diseases during the pre-harvest period. When integrated into CSA, it can aid farmers in ensuring highly effective water management system, reducing wastage, increasing production, improving economic wherewithal of farmers and peasants as well as reducing the impact on the environment. An application called 'SmartFarm', a data-driven farming solution created by Bengaluru-based CropIn Technology, has enabled farmers in the southern part of the state of Punjab (as part of a pilot program). It has created a web and mobile application that enables efficient resource planning, data analytics and modelling in the agriculture sector. It provides information on cropping conditions, the temperature conditions, right proportion chemicals required, issues of crop infection, time for harvest, quality, size and the overall quantity of yield [7].

The Indian-based company Tech Mahindra has developed 'Farm Sensor' and 'Farm to Fork' to measure critical farm attributes at the soil, the crop level and above the soil level. It could record different critical attributes from the field and with the help of big data and analytics predict, prescribe and warn the farmer of the inputs, diseases and weather conditions that help him take accurate mitigation or remediation; right when it is needed []. The multi-stakeholderism model should enable greater representation as well as participation of farming communities, indigenous local communities, women groups, NGOs and independent renowned researchers in the field. A body in that manner should be created by the government at state and district level in both the countries. This provides for greater sustainability in agricultural and allied activities and thus greater inclusive growth at the grassroot level. This also enables the integration of top-down and bottom-up approach, thus enabling greater role for farmers in nation building process.

But the situation in various African countries is complex and different, with agricultural productivity declining proportionally to the increasing conflict among various state(s), non-state actor(s) as well as the nature. According to reports, the South Sudan is in the midst of conflict due to external as well as internal actors that have disrupted trade, prevented humanitarian access to a large extent and affected a large number of livelihoods. The internally displaced persons (IDPs) are increasing in number besides the large number of refugees living in camps on the border areas or neighbouring countries. Besides, the deficit of rainfall in Eastern Equatoria and southern Jonglei has led to reduction in the quality of pasture as well as water availability, affecting the civilians as well as livestock. In Nigeria, the north-eastern region is dominated by the Boko Haram and increased conflict with the government. Besides limiting the cultivable land available, the conflict has led to displacement of nearly two million people as well, besides the already existing 3.5 million refugees, IDPs and returnees. In Ethiopia, severe drought has hit the country, largely affecting the Ethiopia–Somalia region, reducing the income and savings of households and increasing incidences of poverty and famine. Besides, the rainfall has been scant in the Afar region, where below-average vegetation conditions exist. In Somalia, until April 2018, there was severe drought conditions (prolonged) resulting in huge loss of livestock population and production losses. There has also been recurrent conflict in other parts and flash floods affecting the lower Jubba and lower Shabelle, all of which have increased the IDPs. In Democratic Republic of Congo (DRC), the exacerbated conflict in Kasai region, North Kivu, South Kivu, Ituri and Tanganyika Provinces have affected farming communities and increased poverty as well. In southern part of Africa, rainfall has been poor leading to increased drought conditions and lowering of production in countries like Malawi, Mozambique, Madagascar and Zimbabwe [9].

5 Conclusion

The increasing digitalisation that is taking place in the form of IoT, blockchain technology and artificial intelligence (AI) has all created newer paradigms of opportunities and increased potential for transforming traditional forms of agriculture into digital agriculture. The emergence of big data has transformed these efforts to enhance the development of both and enables enhancement of cooperation and collaboration at a greater level. There should be collaboration between government, industry and academia for generating sustainable solutions by harnessing the power of big data and digital innovation. This can help bridge the bottlenecks in ensuring sustainable development of agriculture and allied sectors through efficient and effective support for enhancing agricultural productivity, modernising agriculture as well as in investing for value addition in food processing industries. India should focus on joint research, cooperation and partnership with other nations, both developed, developing and LDCs in the field of big data and CSA. It can vastly transform the field of agriculture and allied activities, logistics, storage, food processing and other related supply chains, even globally. This ensures as well as enables better innovation for the generation of a distinct data analysis. It has a greater possibility of essentially transforming agriculture in India. Now, rather than the usual distinctive character maintained as far as agriculture is considered, the integration of big data particularly into CSA-induced techniques and mechanisms. This can help imbibe the possibility of development not just for farming communities for sustenance, reduce poverty, increase exports to the poverty-stricken underdeveloped regions in the country. This can help meet the goal of national development as well the international targets under Sustainable Development Goals (SDGs) with 2030; particularly that of elimination of poverty, climate change as well as reducing inequality. CSA coupled with big data could help in early warning techniques, ensure pest control of new varieties and optimise production and regulations for quality control.

African countries have the potential not just to become self-sufficient but become a major exporting nation in the global agmarkets. The need to address agricultural productivity, efficiency and development is critical for ensuring food security, nutrition security, livelihood security, climate resilience and ensuing a healthy trade environment within and between nations, particularly for India and Africa. This is critical in regard to the current food crisis fuming in South Sudan, Somalia, Ethiopia and other parts of Africa. The emergence of big data will transform these efforts to enhance the development of both and enables enhancement of cooperation and collaboration at a greater level. The domain of development of big data has been able to interface with different layers of governance level by enabling effective and efficient local governance, agriculture, decentralised planning and participative decision-making through access to big data which has been enabled by low-cost or intermediary technologies. This collaboration is between government, industry and academia for generating sustainable solutions by harnessing the power of big data and digital innovation. This can help bridge the bottlenecks in ensuring sustainable development of agriculture and allied sectors through efficient and effective support for enhancing agricultural productivity, modernising agriculture as well as in investing for value addition in food processing industries.

India should focus on joint research, cooperation and partnership with African countries in the field of big data, as it can vastly transform the field of agriculture, ensure as well as enable better innovation for generation of distinct data analysis. It has a greater possibility of essentially transforming agriculture in both India and African countries. It should also create infrastructure as well as establish a dedicated India–Africa Agriculture Development Fund. Rather than the usual distinctive character maintained as far as agriculture is considered, the integration of big data particularly into CSA-induced techniques and mechanisms imbibes the possibility of development not just for farming communities for sustenance, but also for meeting the greater demand to uplift more than a third of the population from poverty, increase exports to the poverty-stricken underdeveloped regions in Africa, and thereby help meet the goal of national development as well the international targets under Sustainable Development Goals (SDGs) with 2030, particularly that of elimination of poverty, climate change as well as reducing inequality. The future agricultural systems can exploit the ever-increasing quantities of complex, massively multivariate data concerning all aspects of crop determinants, land management practices, probably by domesticating data and ensuring accessibility to individuals in real time. This will not only help in improve decision-making, but increase outcomes and even maximise dividends to different stakeholders to all levels of the society, ensuring economic and social stability within the India and SSA.

Declaration of Conflicting Interests The author declared no potential conflicts of interest with respect to the research, authorship and/or publication of this article.

Funding
The author has not received any financial support for the authorship, research and/or publication of this article.

References

1. Bhargava, Y.: India urgently needs a data security policy: V.C. Gopalratnam. The Hindu (2017). http://www.thehindu.com/business/Industry/india-urgently-needs-a-data-security-policy-vc-gopalratnam/article18072254.ece. Accessed 28 Aug 2017
2. Bordoloi, B.: Big Data for the next green revolution. TheHinduBusinessLine (2017). https://www.thehindubusinessline.com/opinion/big-data-for-the-next-green-revolution/article21850786.ece1. Accessed 10 Sept 2018
3. Bose P.: privacy issues in data analytics: The Indian Perspective (2016). http://www.advayalegal.com/blog/privacy-issues-in-data-analytics-the-indian-perspective/. Accessed 28 Aug 2017
4. Charles, D.: Should farmers give John Deere and Monsanto their data? (2014). http://www.npr.org/sections/thesalt/2014/01/21/264577744/should-farmers-give-john-deere-and-monsanto-their-data. Accessed 1 Sept 2017
5. Ciceri, D., Allanore, A.:. Local fertilizers to achieve food self-sufficiency in Africa. Sci. Total Environ. **648**, 669–680 (2018). https://doi.org/10.1016/j.scitotenv.2018.08.154. Accessed 10 Sept 2018
6. Columbus, L.: 10 charts that will change your perspective of big data's growth (2018). https://www.forbes.com/sites/louiscolumbus/2018/05/23/10-charts-that-will-change-your-perspective-of-big-datas-growth/#445585442926. Accessed 9 Sept 2018
7. Dias, C.: CropIn Technology is digitising Indian Agriculture, One Farm At A Time (2016). http://www.networkedindia.com/2016/12/28/cropin-technology-digitising-indian-agriculture-one-farm-time/.Accessed 25 Nov 2017
8. Ehui, S., Pender, J.: Resource degradation, low agricultural productivity, and poverty in sub-Saharan Africa: pathways out of the spiral. Wiley Online Library **32**(1), 225–242 (2005). https://doi.org/10.1111/j.0169-5150.2004.00026.x. Accessed 15 Sept 2017

9. Famine Early Warning Systems Network (FEWS NET): Areas of highest concern (2018). http://fews.net. Accessed 10 Sept 2018
10. Federal Ministry for Economic Cooperation and Development (BMZ): Federal Ministry for Economic Cooperation and Development (2016). https://www.bmz.de/en/. Accessed 17 Sept 2017
11. Food and Agricultural Organisation of the United Nations (FAO): Climate-Smart Agriculture (2018). http://www.fao.org/climate-smart-agriculture/en/. Accessed 10 Sept 2018
12. Food and Agricultural Organisation of the United Nations (FAO): SUMMARY, Farming Systems and Poverty: IMPROVING FARMERS' LIVELIHOODS IN A CHANGING WORLD. Resource Document (2001). http://www.fao.org/3/a-ac349e.pdf. Accessed 10 Sept 2018
13. Food and Agriculture Organization of the United Nations (FAO), International Fund for Agricultural Development (IFAD), World Food Programme (WFP).: The State of Food Insecurity in the World. Meeting the 2015 international hunger targets: taking stock of uneven progress. Resource Document (2015). http://www.fao.org/3/a-i4646e.pdf. Accessed 9 Sept 2018
14. Gilpin, L.: How Big Data is going to help feed nine billion people By 2050—BI Insight. Business Intelligence (2014).https://businessintelligence.com/bi-insights/big-data-going-help-feed-nine-billion-people-2050/. Accessed 30 Aug 2017
15. Gupta, S.: Data Science in the Indian Agriculture Industry (2018). https://www.analyticsvidhya.com/blog/2018/05/data-analytics-in-the-indian-agriculture-industry/. Accessed 9 Sept 2018
16. Haq, Z.: Farming in India unprofitable for nearly 2 decades. The Hindustan Times (2018). https://www.hindustantimes.com/india-news/farming-in-india-unprofitable-for-nearly-2-decades/story-obZl6IcrjAxIBEjOoF4B6J.html. Accessed 10 Sept 2018
17. Harvey, C.: Big Data challenges. *Datamation* (2017). https://www.datamation.com/big-data/big-data-challenges.html. Accessed 30 July 2017
18. Helbing, D., et al.: Will democracy survive Big Data and artificial intelligence? Scientific American (2017). https://www.scientificamerican.com/article/will-democracy-survive-big-data-and-artificial-intelligence/. Accessed 21 Sept 2017
19. Indian Technical & Economic Cooperation (ITEC).: About ITEC (2015). https://www.itecgoi.in/about.php. Accessed 9 Sept 2018
20. Indo Asian News Service (IANS): Organic farming in India has huge opportunities: Niti Ayog CEO Amitabh Kant (2017). http://www.hindustantimes.com/india-news/organic-farming-in-india-has-huge-opportunities-niti-ayog-ceo-amitabh-kant/story-4KGsyOqsNeB6MrcqXLUhQP.html. Accessed 1 Oct 2017
21. International Data Corporation (IDC): Double-digit growth forecast for the worldwide big data and business analytics market through 2020 led by banking and manufacturing investments. According to IDC (2015). https://www.idc.com/url.do?url=/includes/pdf_download.jsp?containerId=prUS41826116&position=51. Accessed 20 Aug 2017
22. Laney, D.: 3D data management: controlling data volume, velocity, and variety. Resource document (2001). Gartner's blog 2001. http://blogs.gartner.com/doug-laney/files/2012/01/ad949-3D-Data-Management-Controlling-Data-Volume-Velocity-and-Variety.pdf. Accessed 20 Aug 2017
23. Leff, B., Ramankutty, N., Foley, J.A.: Geographic distribution of major crops across the world. Global Biogeochem Cycles **18**(1) (2004). https://doi.org/10.1029/2003GB002108. Accessed 10 Sept 2018
24. Marr, B.: Big Data: Using Smart Big Data, Analytics and Metrics to Make Better Decisions and Improve Performance. Wiley, Chichester (2015)
25. Meola, A.: Why IoT, Big Data & smart farming is the future of agriculture(2016). http://www.onfarm.com/iot-big-data-smart-farming-future-agriculture/. Accessed 2 May 2018
26. Ministry of External Affairs (MEA): Joint press conference following the conclusion of first India-Africa Forum Summit 9th April 2008, Vigyan Bhavan, New Delhi (2008). https://www.mea.gov.in/media-briefings.htm?dtl/4075/Joint+Press+Conference+following+the+conclusion+of+first+IndiaAfrica+Forum+Summit+9th+April+2008+Vigyan+Bhavan+New+Delhi. Accessed 9 Sept 2018

27. Ministry of External Affairs (MEA): Pan African e-Network Project (PAENP). Resource document (2018). https://mea.gov.in/Portal/ForeignRelation/Pan_African_e_docx_for_xp.pdf. Accessed 9 Sept 2018
28. Modgil, S.: Startup watchlist: 12 Indian Agritech startups to watch out for in 2018 (2017). https://inc42.com/features/watchlist-agritech-startups-2018/. Accessed 10 Sept 2018
29. Rioux, J., et al.: Planning, implementing and evaluating climate-smart agriculture in smallholder farming systems (2016). http://www.fao.org/family-farming/detail/en/c/422925/. Accessed 12 Sept 2017
30. Saggi, M.K., Jain, S.: A survey towards an integration of big data analytics to big insights for value-creation. Inform. Process Manage **54**(5), 758–790 (2018). https://doi.org/10.1016/j.ipm.2018.01.010. Accessed 10 Sept 2018
31. Schwab, K.: The fourth industrial revolution: what it means and how to respond. World Economic Forum (2016). https://www.weforum.org/agenda/2016/01/the-fourth-industrial-revolution-what-it-means-and-how-to-respond/. Accessed 18 July 2017
32. Srivastava, M.: Mobile Internet users in India to double by 2017, says study (2015). http://www.livemint.com/Industry/VThUq5I4BivpTDZdQb5sNN/Mobile-Internet-users-in-India-to-double-by-2017-says-study.html. Accessed 25 Sept 2017
33. Steinhart, M.: Infographic: Hadoop & Big-Data Jobs in India (2014). https://www.allanalytics.com/author.asp?section_id=3314&doc_id=271491. Accessed December 20, 2017
34. Stubbs, M.: Report on Big Data in U.S. Agriculture (2016). https://www.everycrsreport.com/reports/R44331.html. Accessed 25 Sept 2017
35. The Economic Times: India and Israel joint statement: Full text. *The Economic Times* (2017). http://economictimes.indiatimes.com/news/politics-and-nation/india-and-israel-joint-statement-full-text/articleshow/59459910.cms. Accessed 8 Sept 2017
36. United Nations (UN).: World population projected to reach 9.8 billion in 2050, and 11.2 billion in 2100. UN DESA Department of Economic and Social Affairs (2017). https://www.un.org/development/desa/en/news/population/world-population-prospects-2017.html. Accessed 18 Sept 2017
37. Wolfert, S., et al.: Big Data in smart farming—a review. Agric. Syst. **153**, 69–80 (2017). https://doi.org/10.1016/j.agsy.2017.01.023. Accessed 10 Sept 2018
38. World Bank: India: issues and priorities for agriculture (2012). http://www.worldbank.org/en/news/feature/2012/05/17/india-agriculture-issues-priorities.Accessed 8 Sept 2017

IoT-Enabled Agricultural System Applications, Challenges and Security Issues

Padmalaya Nayak, Kayiram Kavitha and Ch. Mallikarjuna Rao

Abstract The growing demand of Internet of Things (IoT) brings many more paradigms in several areas of applications such as smart city, smart village, smart energy management, smart agriculture, smart health care, etc. It aims at integrating the virtual world along with the physical world by using the Internet as communication medium. The IoT could be practically feasible with several existing technologies such as wireless sensor network (WSN), radio frequency identification (RFID), middleware technologies, cloud computing and end-user applications. The technologies associated with the IoTs have great impact on precision agriculture or smart farming as well as global economy. This chapter aims at agricultural applications where it utilises modern technologies that benefit the farmers with decision tools and reduces manual labouring cost. The seamless integration of products, knowledge and services through IoT maximises the volume of productivity, product quality and profit of business. Even though current surveys on the IoT in agriculture focuses on the challenges, constraints, benefits and pitfalls for large scale in the agricultural food sector, all are presented in isolation to each other. So, keeping all in these in mind, a brief discussion on challenges, benefits, constraints, future trends and security issues are presented in this book chapter.

Keywords IoT · Smart farming · IoT-enabled architecture

1 Introduction

IoT is one of the promising technologies that connects billions of devices through the Internet and applications of IoT are huge in range from defence to health care,

---

P. Nayak (✉) · K. Kavitha · Ch. Mallikarjuna Rao
Gokaraju Rangaraju Institute of Engineering and Technology, Hyderabad, India
e-mail: padmalaya@griet.ac.in

K. Kavitha
e-mail: kavitha.bits@gmail.com

Ch. Mallikarjuna Rao
e-mail: cmrao@griet.ac.in

P. K. Pattnaik et al. (eds.), *IoT and Analytics for Agriculture*, Studies in Big Data 63,
https://doi.org/10.1007/978-981-13-9177-4_7

classroom coaching, industry, agriculture, etc. [1]. Agricultural production has a vital role in Indian's economy. The GDP of a country mostly grows proportional to the agricultural productions of a country. In India, the share of agriculture is 16% of total GDP and 10% of total exports. The overall GDP of the country has been decreased due to the share of agriculture compared with the other sectors. Countries like India and China need to increase their agriculture production due to their fast-increased population growth. Due to the increasing population in the world, the importance of agriculture is quite natural. According to the report by United Nations of Food and Agriculture Organization, it is expected that the population will be doubled by 2050. The increased production of the agriculture through modern technology using IoT will support huge economic lift to India. In agriculture, rice is the major dominating product compared to other crops and majority of the people in the world love to consume rice. Specifically, around 70% of the people in India consume rice as their major food. India's rank is the second largest producer of rice after china across the globe. The farmers face lot of difficulties in the crops due to the pests and diseases in crop, managing the soil, water, climatic change, etc. [2]. Mostly crop losses occur up to 37% due to pests and diseases every year. There are many diseases like fungi, bacteria, viruses, nematode caused by rice and it occurs frequently where ever rice is grown. Some of them are national and international importance, others occur in local areas. These crops-related problems can be resolved with the IoT-based user friendly intelligent systems. Nowadays, Internet of Things (IoTs)-based technology is going to change the structure of agriculture/industry sector and connects the stakeholders to contend with the challenges they face. The quantity and quality of the crop production can be enhanced by addressing all these issues. The advancement in wireless communication systems and VLSI design has led to the development of smart tiny sensors, handy mobile devices and ubiquitous services through Internet connectivity. It has created a new era of IoT-based agricultural applications. According to Gartner's prediction, there will be 25 billion IoT devices in the market by 2020.

According to the research report produced by Machina, the number of agricultural devices through Internet connectivity is predicted to raise from 13 million to 225 million from 2014 to 2024 [3]. The Internet Protocol (IP)is the heart of network layer that provides the route to carry the data, exchange the data and network-related information between multiple LANs called as distributed system. The connected IoT devices via Internet make the distributed system to enable the data to be accessible any place, anywhere and at any time. However, the transfer of data over a huge network through Internet services requires adequate security, privacy, data management attention as Internet Protocol (IP) is subject to theft analysis and not enough security features are inbuilt to it. The Internet with cloud computing technology has further provided a huge platform where large volume of data can be gathered, processed, analysed and accessible to the stakeholders. To manage cloud services is another challenging task [4] where many factors are involved like middleware's, APIs, services, establishing and managing the network nodes and dealing with the data in a proper computational speed. In addition, the connectivity of heterogeneous systems and devices over the Internet, communication algorithms/protocols and IoT middleware are the main building blocks for any type of application. Service-oriented architecture (SOA) is the example of middleware of IoT which consists of multilayer architecture. In extension, cloud-based IoT middleware and actor-based IoT middleware are discussed in [5, 6]. In few proposed architectures, how the data can be sensed, fetched in a connected network, how to handle middleware and application layers are discussed [5–8]. Interested researchers can refer these literatures.

1.1 IoT-Enabled Technologies Used for Smart Farming

IoT-based agriculture is about to empower farmers with automation tools, decision power and automated technologies that integrates information, service and product to maximise the product and minimise the waste. This section gives an outlook of IoT-based smart farming where modern technology can improve the farmer's life, enhance the productivity, minimise the cost and extend the supply chain as shown in Fig. 1. There are three layers active in IoT structure. These are perception layer, network layer and application layer. In perception layer, we meet sensors, RFID technologies, near-field communication (NFC), etc. In second layer (network layer) where the sensor nodes interact with the environment, communicate with the neighbour nodes or gateways, build the network by forwarding the data and send the data to the remote

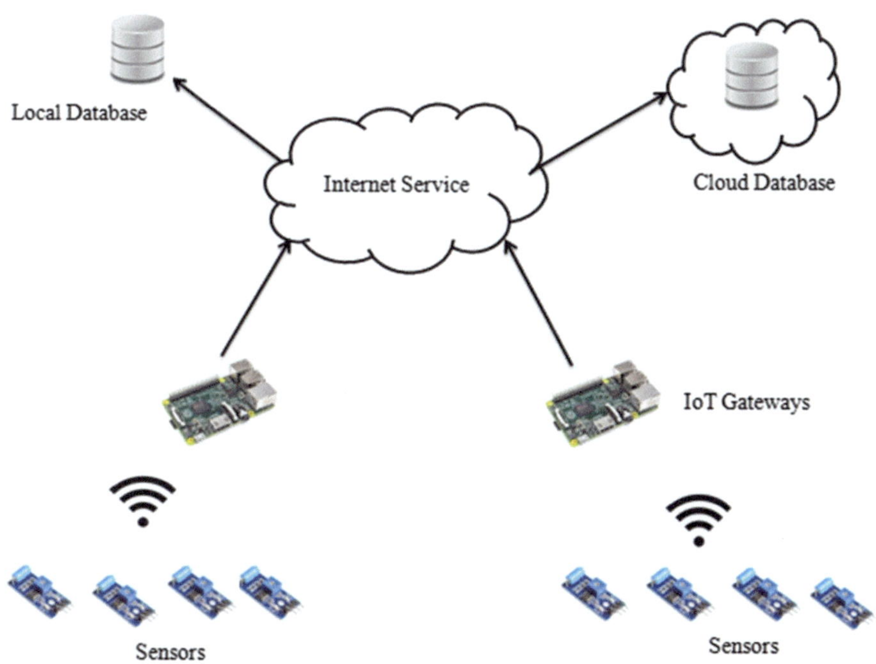

Fig. 1 Example of IoT-based farming

server for further processing. Third layer, application layer, faces many challenges and issues which has to be resolved. Application layer must provide the access to the authenticated users and modify the data. Every device should be identified with unique IP address. To provide billions of devices unique ID and address across the globe is not an easy task and each device must be empowered through Internet access and control over it. Future Internet Protocol will take care of it. In real-time scenario, sensors usually do their miracle by performing sensing activities. The major part of smart farming involves the deployment of sensors in outdoor environment which can sense the various parameters like temperature, humidity, pressure, moisture in the soil, crop's growth rate, cattle health, chicken production, etc. This large volume of sensed data are streamed via wireless link to the IoT gateway, to the local server with a replica stored in the cloud server. This big data is processed, analysed and available at the cloud so that stakeholders can access the data at any time and take care of the crop products, live stocks, etc. This reduces the production cost dramatically. Also, with livestock monitoring, you can mitigate risks of losing the yield. In addition, the waste management can be done more efficiently for many by-products. This can help in better cost management and increased revenue generation. Such process automation opens up many avenues for pest control, fertiliser recommendation with enhanced product quality and volumes.

2 Applications of IoT in Agriculture

Over the decade, we have witnessed the evolution of Internet into wireless devices providing the devices a unique address on the network. The main pillar of IoT technology is IPv6 (instead of IPv4) that provides a unique address to each object in this world. These devices are connected via Internet and have many popular applications and agricultural production is one such application. Now-a-days, IoT-based smart living, smart villages, smart city, smart machinery are high in demand. As agriculture is back bone for many countries, modern tools and technology must be utilised for proper food supplies across the world. Several environmental factors such as temperature, humidity, salinity, leaf wetness, soil moisture, dry circle, solar radiation, rainfall, pest movement and human activities can improve the agricultural productivity. If the farmers can able to monitor these factors remotely using smart technology, surely our County's economic condition can be increased. Farmers can take some preventive measures for pest control based on historical information regarding pest control, get a chance to improve the quality of farming and achieve high productivity. There are several applications of IoT-enabled agricultural system which are listed below. Figure 2 depicts the applications of IoT-enabled agriculture.

- Monitoring activities
  - Monitoring irrigation and water quality
  - Monitoring weather condition
  - Monitoring soil

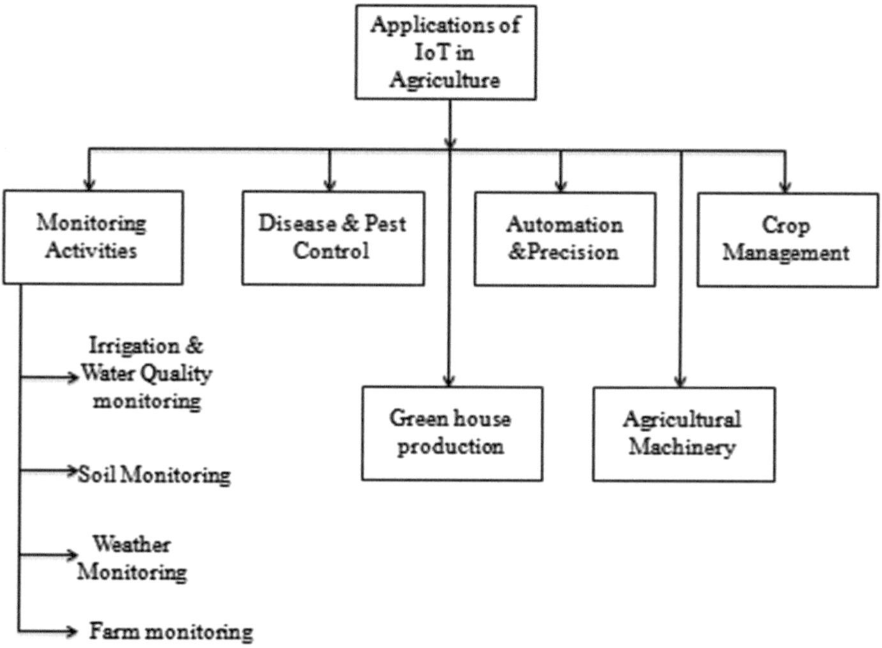

Fig. 2 Applications of IoT in agricultural sector

- Monitoring the farm

- Crop management
- Agriculture machinery
- Automation and precision
- Disease and pest control
- Greenhouse production.

2.1 *Monitoring Activities*

Today, the most critical task for farmer is to monitor, model and manage the environmental parameters which are beyond predictions by engineers and scientists. The Agricultural activities require continuous monitoring for accurate and reliable advice from agronomists. There are many parameters which need to be monitored such as environment, cropland fertility, water quality for aqua life. In this section, we present overall view about the use of IoT in monitoring system for weather, soil, water and smart farms.

2.1.1 Irrigation and Water Quality Management

Irrigation plays a major role to achieve sustainability in agriculture. Automated irrigation system [9] is the primary requirement in agriculture. So, wireless sensor nodes are deployed across the agricultural area to sense various parameters required for irrigation. The soil moisture sensor and temperature sensors collect the respective readings from the home gardening or agricultural area and push the sensed data into the cloud. The farmer can access the streamed data at any place and any time. Crop sensing for various parameters is done accurately by using IoT technology. Remote monitoring can be done to monitor and control the agricultural field using wireless senor devices. To monitor soil quality and ground water level in agricultural field, ultrasonic waves have been sed. The use of IoT-based (using Arduino) smart drip irrigation system [10] aids in optimal irrigation facility for crops. Water distribution is a major challenge even today in many developing countries especially in water scarce regions. There is a need for automatic water monitoring and distribution system to enable uniform distribution of water among the stake holders. Smart water metres need to be deployed on a large scale. The system proposed in [11] uses smart water metering in the Indus basin. There is also a need of IoT-based technology to enable demand-based water deliveries. This system has put efforts in automating the large-scale irrigation networks.

A real-time water quality monitoring at a low cost is developed for IoT environment. The use of IoT has revolutionised as the innovative technology has connected to web. Irrigation is the primary resource for agriculture. The WSNs are widely used in irrigation management applications.

2.1.2 Monitoring the Weather

The major climatic factors affecting the agriculture are hail, frost, snowfall, flood, drought and storms. So, a weather prediction system is highly desirable, which can monitor parameters like air, temperature, humidity, barometric pressure, light intensity, rain, speed and direction of wind. So that, farmer can properly plan for sowing seeds, fertilisation of plants, sub-soiling, ploughing, irrigation, etc. activities based on the weather conditions like sunny, rainy, cloudy, snowing, etc. This helps in better utilisation of resources at a proper time. The low range (LoRa), low power consuming devices is used to check the feasibility and economy for farmers in weather monitoring. The system has been demonstrated on a wide agricultural field with several sensors deployed to collect the physical parameters. These readings are relayed to the weather station, and the data is available through a web application. This system was connected through LPWAN through LoRa-based RF transceivers. During the development of hardware prototypes, Arduino nanoboard is used as an open-source electronic prototyping platform, which consists of an Atmel microcontroller mounted on a circuit board. The farmer can view the web application for up-to-date information about the weather conditions without visiting the field. Nowadays, the weather monitoring systems exhibit many capabilities. The wireless sensor network

(WSN) is perfect choice for such monitoring applications. As the sensor nodes can deploy themselves in remote environment, they are able to reorganise themselves even after node failures. The IoT management system can protect the balance in ecology and improve the quality of environmental parameters. A case study is proposed in [12] which integrates geo-informatics, cloud computing and information systems for accurate monitoring and management of environment.

In this system, multiple sensors are used to collect data and web services to transfer that information to the server. This information system resulted in benefit of not only in data collection giving valuable big data analytics, but also the monitoring and decision-making improved a lot through the web services and cloud computing platforms.

2.1.3 Monitoring the Soil

One of the primary factors to determine plant growth is moisture in the soil. This plays a vital role in providing the essential nutrients to the growth of a plant. To determine moisture in the soil, resistive sensor can be employed. These sensors determine the change in soil resistance between two probes of the soil moisture sensor. Due to global warming, the climatic conditions have become unpredictable. The amount of rainfall is decreasing every year leading to scarcity in water and essential soil nutrients. This necessitated for automated irrigation system [9] to determine the macronutrients, i.e. nitrogen, phosphorus and potassium in the soil. This system can save time, money and power for the farmer. Many farmers are using manual methods to perform this test, which require repeated human intervention and is a tedious process. With this automated mechanism, farmer is able to receive the soil fertility and the irrigation information via e-mail directly. The IoT-enabled sensors are deployed in the agricultural land to capture the details about soil and climate.

The soil moisture sensor gives indication of water content in the soil. The growth of the plant can be determined by measuring the moisture contents in the soil. Whenever water content goes beyond the threshold, an alert message is generated to supply water in that area. These automatic intelligent devices are very much helpful in remote monitoring using IoT. In case of inadequate nutrients in soil, farmers use fertilisers to make the soil fertile. In real time, IoT monitoring system can be used to identify the requirement of soil and suggest the amount of fertiliser required. This will maintain the health of the soil to aid for plant growth. In this effect, the applications of IoT are huge in agricultural field that can be applied monitoring weather condition, environmental condition, controlling greenhouse automatically, industrial intelligent control system, tracing food safety measure, etc.

Apart from these, temperature and humidity are few major environmental parameters that affect weather change, topological change, types of soil and vegetation [13]. In this way, the temperature sensor gives measure about the coolness and heat content variation in the soil. Some of the agricultural robots also determine the dry lands while weeding the crops and suggest the irrigation requirements to the farmer.

2.1.4 Monitoring the Farm

There are many innovative applications of IoT [14–16] for connected farm management systems for agricultural industry. There are several farming activities which need attention for innovation and better management. Some of the widely found farming industries are cattle farms, bird farms, beehives, etc. The use of IoT in these farms helps the farmer in achieving higher production. Beehives are used not only for honey production, but also there is a fact that beehives greatly aid pollination of crops. The farmers have to regularly monitor the beehive status. So, a beehive monitoring system [17] gathers internal and external data of the hive. These data determine the position of the bee colony from a set of data using a classification algorithm based on decision trees. Farmers can directly take a decision for his visit to the hive, based on the sensed data available on his smart phone. Also, the study resulted in prediction of rain forecast based on the beehive status. In addition, smart beehive provides automatic honey production through its automated machinery. So, it reduces the human risk of squeezing the honey from the beehive. This beehive too can greatly help farmers in agriculture production. Some of the smart farming techniques are displayed in Fig. 3.

As the world population is increasing at a rapid rate, there is a high demand for food production. The poultry industry is in critical situation unable to cater for growing food needs. The use of efficient farming techniques and some best practices for improving the production can solve the problem. The study conducted in [18] has

Fig. 3 Example of smart farming

pointed out the use of technology for productive management. It explores the use of smart phones to monitor chicken farm activities.

The poultry industry is in need of various monitoring services rendered by IoT. One such activity can be seen in the hatcheries of poultry industry. Traditionally, the birds give the required temperature for 21 days to hatch the eggs. But due to various reasons, some of the eggs are not hatched while some remain incomplete. In a huge poultry, it is extremely impossible to provide every bird with all the ideal characteristics. This severely hampers the profits for the poultry farmers. The IoT-based hatcheries come to farmers rescue. Such system can provide the required temperature and humidity for hatching the eggs. This improves the hatch rate, production, chicken quality and operational efficiency. Also, the monitoring service of the hatchery can help the farmer in detecting health issues in chicken. It is very difficult to identify the infected chicken for a long time until the disease spreads to other chicken and there is huge loss to the farmer. Using IoT, these birds can be monitored continuously and the respective temperature and well being of the chicken can be reported to any remote location with these IoT devices. These data can instantly be rendered on smart phone. In another instance, the chicken farmer monitors the poultry using smart lighting system [19]. The smart lighting system can be controlled by adjusting the lighting intensity to generate the required temperature inside the poultry using a smart phone. In addition, there is mechanisation of certain activities of feed and water supply to these egg laying birds at regular intervals of time aids in replacing the human intervention for accuracy and timely service.

Another important smart farming applications can be seen in cattle or animal farms. Different animals like cow, buffalo, goat, sheep etc. farm animals not only contribute to meat industry but also their byproducts like milk, wool, dung and organic manure for plants are in high demand. With the use of IoT devices, many of the traditional farming methods are replaced with artificial intelligence-enabled IoT devices and robots. Besides regular animal movement, these IoT devices can also monitor the health and fertility of the cattle. The fertile cycle for the cow is a critical 8 h window in one month of time. So, the Moo monitor can continuously monitor the cow from a distance of 1000 metres can help the farmer in detecting the exact time window to help the cow for reproduction. A neck-mounted device worn by the cattle whenever the cattle cross a defined Wi-Fi point, the information is transmitted to the mobile device of the farmer via cloud. This helps in monitoring the health of the cattle remotely to take necessary precautionary measures in case of birthing and maternity.

The cattle milking robots are in high use in many advanced countries. These robots not only replace manual milking activity, but also take care of the health of the cattle. The cattle are tracked with a unique RFID and monitored for their milking session. The robot gets the information when the animal is ready to be milked. Also, their health is monitored and fed with special grains during lactation. The giant milk industry is completely dependent on solutions provided by IoT and artificial intelligence. Thus, IoT is the main player in the smart diary industry catering to the ever-growing population.

Aquaponics

Aquaponics is one important technique of farming where plants are not grown in soil. They are grown in water using light as their source of photosynthesis and water for the essential nutrients. The continuous water quality monitoring for fish farming in tanks, along with soilless plants is termed as aquaponics [20]. The fish provides natural fertiliser for plant growth and the plants purify the water for fish cultivation. The fish wastes aid to provide the plants with essential nutrients. For monitoring the water quality and various other parameters such as sunlight, salinity and pH level, IoT devices are widely used. Hence, automation is highly desired in such applications, where there is less human intervention and monitoring can be done remotely.

Forestry

Forestry has great impact to produce carbon cycle. An early detection and alert system can protect against many activities such as forest fire by nursing the soil, monitoring the air, temperature, humidity and different types of gases such as carbon monoxide, carbon dioxide, oxygen, hydrogen, methane, toluene, isobutane, ammonia, ethanol, hydrogen sulphide and nitrogen dioxide.

2.2 Crop Management

Agricultural process does not follow only sowing, watering, adding fertiliser, spraying pesticides and collecting harvest from the field but also requires a lot of analytics in every phase of life. For instance, soil moisture, and damp weather condition is required for germination of seeds. So, weather prediction system is desired during germination phase. Similarly, to grow a healthy plant, fertiliser is required. The amount of fertiliser required or pesticide required requires an in-depth analysis for better management of crop. Farmers can take the decisions by maintaining some historical data or with the help of IoT-based management technique. In real time, the information can be accessed or monitored remotely through mobile phones, URLs, etc.

The distributed systems across the greenhouses can also be managed in an efficient way. It provides the users of the applications with various authentication levels. The real-time information can be delivered through e-mail alerts or through SMS in the smart farming technology. SmartFarm Net [21] is a data acquisition, analysis and visualisation platform for IoT-based smart farming applications. It is the largest in the world to provide crop performance analysis and recommendations. This platform can virtually support any type of IoT device and provide data visualisation with zero programming effort. It provides tools for fast and scalable data access from millions of sensor nodes for monitoring crops. The advantages of IoT-based technology promises better infield monitoring of cropland. Finally, smart agriculture is the new weapon against the increasing climate change and global warming. It also increases the crop productivity by efficiently using IoT along with cloud service for an economically feasible solution for farmers.

2.3 Agricultural Machinery

The use of IoT in agricultural machinery, the crop productivity can be greatly improved. Some of the machinery such as unmanned aerial vehicles and robots operated in autopilot mode are the direct applications of IoT in agricultural machinery. The agricultural equipment manufacturer CLAAS [22] has implemented IoT in their equipment. The Precision hawk's UAV sensors can provide information such as wind speed and air pressure. There is a mechanism called sub-soiling which can enhance permeability of soil, improve soil physio-chemical properties, grow crop root system environments without disturbing soil layer structure and facilitate increases in crop yield. This requires the use of agricultural machinery for ploughing and seeding. The government has to monitor the geographical region to identify number of such agricultural machinery being used. An IoT-based sub-soiling system for agricultural machinery monitoring was developed in [23]. This scheme uses convolution neural network-based automatic image recognition system to identify the agricultural machinery spread over a geographical region. There was a study conducted in which examines the productivity of swine farms on usage of IoT-based electronic sow feeder.

In the era of IoT, most of the ancient farming methods are going to be replaced by IoT-based modern technology. The advancement of wireless technology, sensor network along with cloud computing and fog computing has brought a new era in agriculture. It all starts with sensing and monitoring with the help of the IoT-enabled sensors from the remote locations and then transferred directly to the cloud for direct access across the globe. In monitoring the crop growth and crop diseases, IoT-enabled agricultural drones are playing a major role. The drones are unmanned aerial vehicles operated with a remote control for better crop management. Drones can be used for soil analysis which can help in monitoring the nitrogen content (main requirement for germination) in the soil. Also, these drones can be used in sowing of seeds and spraying of fertilisers and pesticides with uniform coverage of the geography. This greatly reduces the cost and aids in better crop management. Drones are fitted with hyper-spectral, multi-spectral and thermal sensors. This helps us the farmers to calculate the crop index and proper planning of crop. Overall, drones replace the human intervention in continuous crop monitoring activities. Example of smart agricultural machinery is shown in Fig. 4.

On the other hand, there are many other options for such video geography monitoring with driverless tractor and robotic vehicles. The manual methods of sowing the seeds is very labourious and cumbersome and often inaccurate. Often the seeds fall out of the guided line for germination. So, driverless tractor used for ploughing and seeding can save farmers time and effort. With the precision seeding devices,

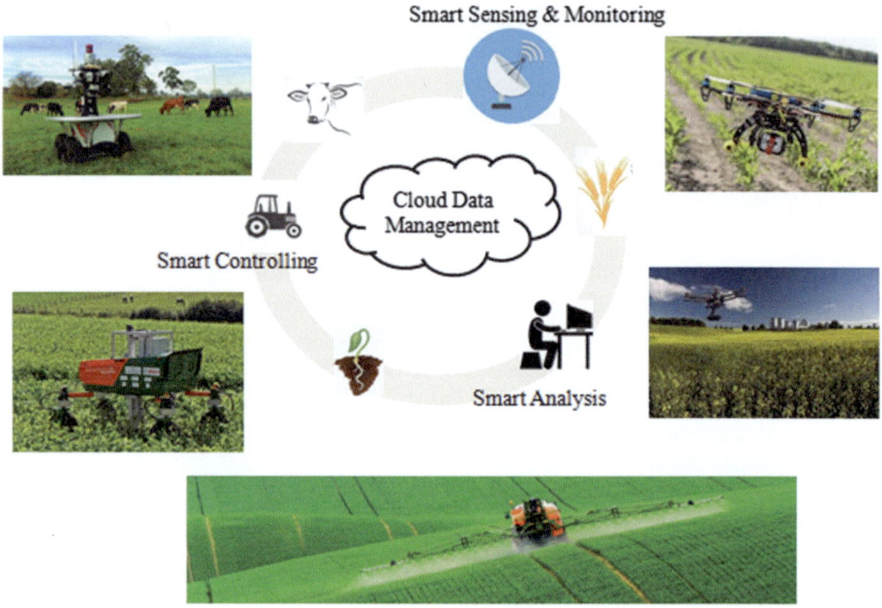

Fig. 4 Example of smart machinery in IoT-based agriculture

sowing and spacing of plants can be done with accuracy without any wastage. This facilitates optimal plant growth. In general, robots can replace human activities, and the precision agriculture can be greatly benefitted by Bonirob robot. It is useful in the weeding of crop. The robot was developed as a prototype in UC Davis smart farm initiative. The robot was developed by integrating artificial intelligence and machine learning. The size of the robot is like a car. It is well trained and can autonomously move around the field and look for weed plants to eliminate. Also, these robots are used in apple plucking kind of tedious human work. It can move around the fields between the plant spacing without damaging the crop.

2.4 Precision Agriculture

Precision agriculture deals with the evolution of traditional technology for miniaturisation of the electronics industry. The precision agriculture is the management of information specific to agriculture to improve the crop production. These include metrological factors (temperature, humidity, sunshine, wind and water), smart irrigation details, etc. which are crucial for farmer in planning his activities and make it profitable. Hence, use of IoT cloud platform [24] offers users massive information collected at a low cost using cloud service. The IoT system implements a decision support system to estimate the water needed for the soil conditions. Also, the farmers can get alert message notified by SMS when the crop needs irrigation facility. Precision agriculture came into existence to optimise the crop production, to improve the crop quality and efficient use of resources like water, power, soil fertility, etc. But, due to the high cost, operation and maintenance of equipment, these networks are not being widely used. The evolution of IoT and embedded systems have allowed the development of economically feasible systems which are easy to install and easy to maintain with minimal power.

2.5 *Disease and Pest Control*

Since several years farmers have been facing difficulties in crop monitoring and disease prediction. Today with the evolution of IoT, farmers are able to achieve good crop production. This section gives details about the usage of IoT in crop monitoring and disease prediction. Disease and pest control is one of the major challenges in agriculture. Every year farmers are losing an average of 37% loss in their crop production due to diseases and pest. With the use of IoT, early disease detection is possible which can facilitate in taking some preventive measures. The continuous crop monitoring activity can be assigned to an IP camera and the recorded video of the crop can be rendered daily to a smart phone. This data could be shared to expert advices to choose a proper pesticide. Also, the farmers need not visit and invite agronomists to the rural areas for their advice on pesticides. Obviously, this kind of monitoring activity could minimise the human intervention, save time and effort. By using deep learning application for disease detection, farmers can classify the disease and also get recommendation of proper pesticide in the smart phone. As smart phones have become affordable, it can change farmer life tremendously with ever-growing IoT technology. For crop disease detection, image processing and for pest control, deep learning algorithms can be used. But deep learning and image processing technologies are limited due to their computational intensity to run on a smart phone. However, very few researches have been carried out in this direction with some limited disease detection for rice crop. The authors in [16] have used inception V3 transfer learning to train on a deep convolutional neural network to classify three types of diseases and two types of pest for cassava leaves. This can be rendered to a smart phone for taking necessary action.

Potato late blight is a dangerous disease for potato crop. There has been a steep decrease in the potato crop quality and productivity due to this dreadful disease. An IoT-based system has been developed to predict the disease based on weather data like temperature, humidity, pressure, etc. and notifies the farmers by SMS to take appropriate measures. The system uses a cloud server to store the weather data like temperature, humidity, pressure, etc. and these data greatly help to the farmers. The wheat disease classification by using smart phone was presented in [25]. The algorithm was limited to identify (rust, septoria and tan spot) on wheat images. They have used smart phone to capture the images and identify the affected region of the plant. Further, the images are processed with a high-level classifier. The disease detection using smart phone is made feasible by reducing the training data size to 100 photos in [26]. This presented a windows phone application for plant disease recognition. This technique uses image processing to analyse the spot colour features on the plant parts. It achieves 90 per cent accuracy when tested on grape diseases. One interesting study was done in [27], which can monitor water level, soil, temperature, humidity, etc. remotely monitored through smart phone or PC. These data are used to study the environment for rice cultivation. The system proposed in [28] has developed a framework to detect plant disease and provide diagnosis in consultation with the agricultural advisory call centre. The plant images are captured by a camera in the field and transfers it an IoT web repository. The system processes the images and provides diagnosis and recommendation for the disease.

2.6 Green House Production

Many farmers use wireless sensor network (WSN) [29–31] for rose gardens to maintain the temperature inside the greenhouse. The WSN is employed to detect the environmental conditions and transmit the data to web or a mobile application to perform data mining and obtain a prediction model with good accuracy. This gives the farmers some relief from monitoring activity in the green house. He can get the status and prediction directly to the mobile application accurately.

An automatic temperature control inside the green house [32] can be achieved using a WSN. The WSN prototype has employed MicaZ nodes to measure temperature, light, pressure and humidity and to share with greenhouse farmers. Greenhouse monitoring using IoT has been very much useful to monitor the light intensity on the plants in green house. This improves the crops production and further maximised the photosynthesis activity. There are many studies for remote monitoring of green houses for their water requirement, soil nutrients, etc. can be done in an effective way.

3 Challenges and Issues in IoT-Enabled Agriculture

This section describes most of the challenges and issues involved in IoT-enabled agriculture. Figure 5 represents the graphical model of challenges and issues involved in IoT-enabled agriculture.

3.1 IoT Hardware and Software Issues

When smart agriculture comes to picture, IoT and some electronic devices immediately come to our mind. Which types of devices can be used for smart agriculture?

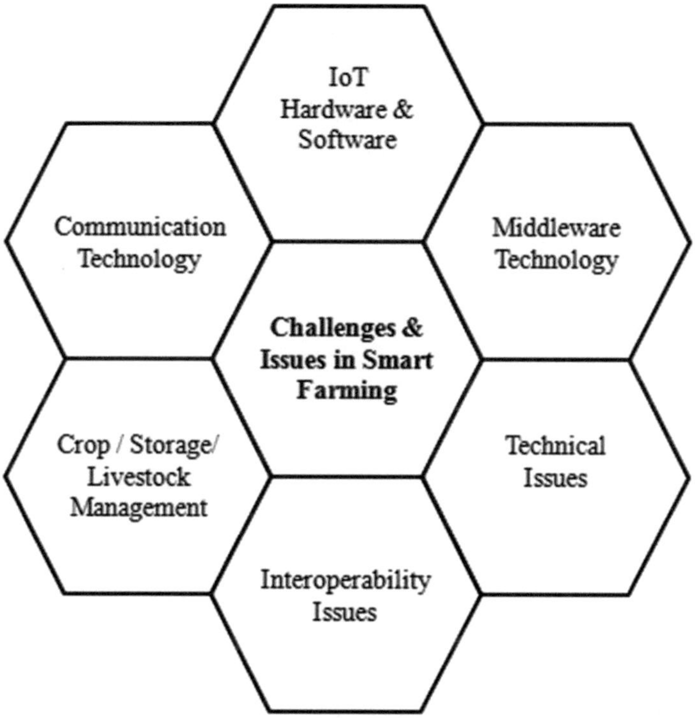

Fig. 5 Challenges and issues in smart farming

These devices must be robust enough to protect them from solar radiation, temperature fluctuation, heavy winds, high rain falls, humidity, all types of environmental changes and other danger activities that destroy the electronic devices. These devices must be battery powered and should actively work for long period of time with limited battery sources. Power harvesting modules like solar panels and turbines can be helpful to some extent for implementing IoT. The devices must be connected to each other through Internet facility which produce large amount of data. This large amount of data (big data) must be handled by small-scale infrastructure of server and the programming tools must be low-power capabilities to be compatible with hardware devices. Further, the IoT devices are required to be deployed in outdoor environment that will expose the devices to frequent changing environmental conditions which may affect the performance of deployed sensors and cause link failures. Basic issues involved in hardware/software are depicted in Fig. 6.

- **Reliability Issue**: There is a big question of physical safety of the deployed IoT sensors and interconnectivity of the networked systems so that data must be reached to the destination. Physical protection must be provided to secure the costly equipment from different climatic changes as well as natural calamities.
- **Scalability Issue**: As billions of IoT devices are required to be deployed in IoT agricultural platform, existing gateways and protocols may not support huge num-

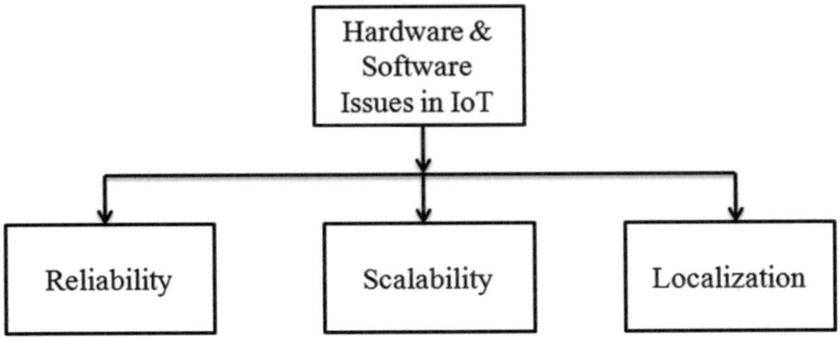

Fig. 6 Hardware/Software issues in IoT-based agriculture

ber of IoT devices/nodes. For instance, Sigfox gateway can support 106 nodes, Ingenu gateway can support 104 nodes, and LoRa supports up to 104 nodes respectively [33]. It requires intelligent/robust IoT management system for each node with a unique address and identification number.

- **Localisation**: When the IoT devices are deployed, several factors are required to be considered. The position of the device must be considered so that it can communicate with other IoT devices and gateways. Its best position, coverage area so that it can provide reliable communicate with other devices without causing any minimal interference.

3.2 Interoperability Issues

Interoperability is a key concept in implementing IoT technology. As millions of devices need to communicate with each other, interoperability is the main crucial and defining element in IoT. There are many ongoing works on protocols and standards needed for billions of devices to interoperate with each other. There are different types of interoperability issues need to be handled in IoT such as technical interoperability, syntactical operability and organisational interoperability. These operability issues are depicted in Fig. 7.

- **Technical interoperability** [21]: Technical interoperability defines the integration of hardware devices with software programming. The programming tools must be lenient enough to be compatible with robust hardware devices.
- **Syntactical interoperability** [34]: Syntactical interoperability is associated with data formats. The syntax of the message must be defined in order to exchange the data from one device to other device (D2D) communication in the form of bit tables or high-level languages like HTML, XML, Java Script Object Notation (JSON), comma separated variables (CSV), electronic data interchange (EDI) as data standard syntax to share the information. Semantic interoperability should well match with users for interchange of information exchange between end-users.

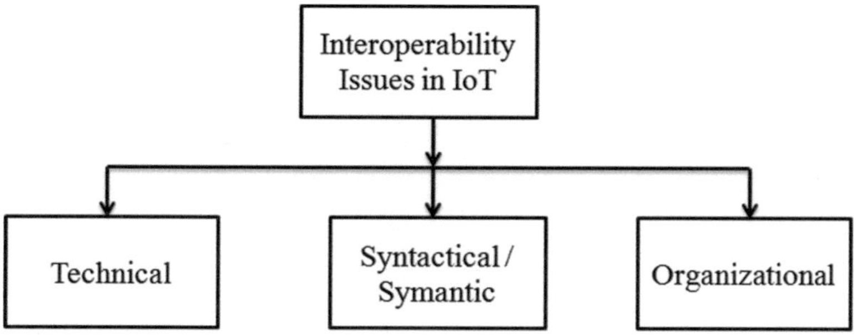

Fig. 7 Interoperability issues in IoT

- **Organisational interoperability** [35]: The effective communication and transfer of meaningful data over highly fluctuating systems and over various geographical regions can lead to a successful distributed system and global-IoT infrastructures.

To handle all types of interoperability issue, open connectivity [36] must be promoted. Further, the usage of adapter services such as If This Then That (IFTTT) [37] allows users to create powerful connections through simple conditional statements are currently being used.

3.3 Middleware Technology

Middleware lies on the top of the network and down of application layers that aim to facilitate interoperability. It takes care for both data processing and communication interface between these two layers. It is basically a connecting software that provides a mechanism for process to interact with other processes running on multiple networked devices.

3.4 Communication Technology

The communication technology has great impact on successful deployment of IoT applications. Wireless communication is the most vital part in IoT applications as the wireless link gets affected very often by the climatic changes, weather conditions, etc. The environmental characteristics not only bring challenges on the hardware devices, but also on the network layer. Basically, wireless communication can be categorised in three ways: based on spectrum, standard and applications. Further, it can be divided into short-range communication or long-range communication. Radio spectrum is again divided into two parts such as licensed and unlicensed frequency spectrum. The unlicensed spectrum utilises 2.4 GHz industrial, scientific and medical band (ISM band). The major drawbacks of using unlicensed frequency band is less

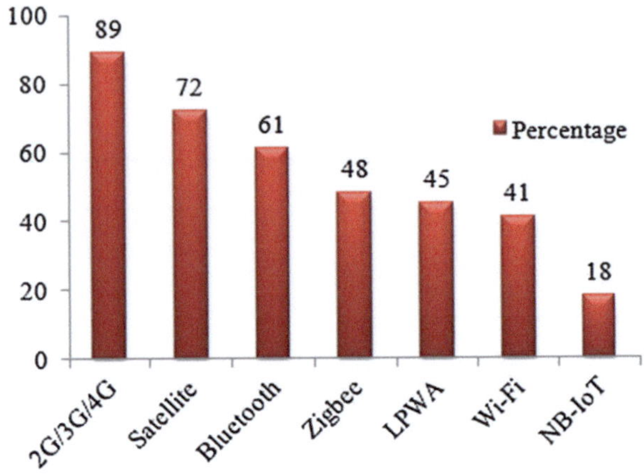

Fig. 8 Communication technologies used in IoT [Ref. from HUAWAI report]

secure, more interference and unreasonable deployment cost. On the other hand, the licensed frequency which is used by cellular communication (GSM network) has less interference, good traffic management, good congestion control management, more reliable, more secure and more quality of service. Despite huge positive impacts, there is a trade-off between cost and power consumption of IoT devices. The wireless technologies used in IoT domain are Groupe Special Module (GSM) authorised by license frequency, wireless personal area network (WPAN), wireless regional area networks (WRAN), mesh network, point-to-point (P2P) and low-power wide-area network (LPWAN). The communication technology for IoT device also depends on type of applications and types of topology to be used. According to HUAWAI marketing report, the usage percentage of all wireless technologies is shown in Fig. 8. Simple network management protocol (SNMP) plays a vital role to manage all the objects, devices and to support communication between them.

3.5 Issues in Crop Management

In crop farming, several factors that affect the farm product includes amount of rain fall, leaf wetness, temperature, humidity, soil moisture, climate, salinity, solar radiation, dry circle, pest movements, human activities, etc. [38]. Those data like type of soil, crop type to be planted and the appropriate pesticide and fertiliser can be structured as a data set. In future, the type of pesticide to be sprayed across the field based on the crop can be identified in advance to save the plants.

3.6 Livestock Management

In livestock management, which factor must be monitored depends upon the types of animals. For example, by experimenting the milk from the buffaloes and cows, health condition of animals can be predicted. According to USDA, farmers in the USA lose nearly $2.4 billion per year due to the sickness of animals and their death. IoT solutions can help farmers to closely monitor the health condition of their livestock, which is an effective way to prevent death rate. Generally, IoT solutions involve sensors attached with the wearables that can be connected to a gateway through low cost, low bandwidth to stream the data to the cloud. The sensors fitted with the wearable of animals can monitor blood pressure, body temperature, digestion, heart rate, respiratory rate and other vital parameters that gives an alarm to the farmer to predict the health condition of animals so that the farmer can take care of the feed issue and illness without waiting for a strong veterinary care. These sensors continuously measure each animal's health, behaviour and send message to farmer, so that farmer can take corrective measure without spending money for the illness. For instance, a cow's temperature rise triggers an alert well before the farmer notices a change in behaviour.

Even to prevent livestock theft, animals are attached with radio frequency identifiers (RFIDs) that can track the animal's movement [39]. The movement of the animal can be visualised on a map in a centralised control system through wireless connectivity. Mostly in rural areas where animals go for communal grazing, usually, animals get lost. RFID chips fitted with wearable really can help farmers to identify their livestock and RFID readers are placed at various monitoring spots to transmit information to an agricultural extension services centre.

3.7 Storage Management

Storage management is an essential activity to maintain food quality and maximise business profit. A large quantity of agricultural product usually gets lost due to lack of proper storage management. The use of IoT can help storage management in an effective way. The contamination of food products is greatly affected by temperature, moisture and humidity, whereas quality of product is affected by insects, rodents, microorganism, etc. Sensors can be deployed to monitor the environmental conditions as well as storage facilities and send to the gateway node. A self-organised automated feedback system must be deployed which can take the decision by analysing the data and adjust the environmental parameters [40, 41]. It can send a warning message to farmers if any deviation occurs in the environment or any pest is reported in the storage facility. In India, many reports show that almost 40% of the fresh product gets spoiled due to pest or spoilage. In [42], a cold storage management based on IoT is designed where the temperature is controlled automatically. Although IoT can manage crop storage efficiently, security is major concern. It can be embedded with storage management system.

3.8 Technical Issues

- **Interference**
  The biggest challenge in the wireless technology is the interference that causes among the IoT devices operating in the same frequency band (For instance, Bluetooth, ZigBee and Wi-Fi) or in neighbouring bands. For IoT application, wireless communication is the key medium that focuses on range of communication, distance, data rate, battery life, mobility, latency, security, resilience, and the cost of gateways and modems. Among the communication technology, the LPWA is attracting so much interest [43] especially with the emergence of narrow band IoT (NB-IoT). The NB-IoT promises interesting features which include low device power consumption, ultra-low device cost, simpler to implement, support of a massive number of low throughput devices, long distance coverage and can support upload and download of data [44].
- **Security and Privacy**
  There are several security issues that need to be addressed during implementation of IoT-based farming. The lack of adequate security may lead to data loss, access of raw information, destruction of sensor nodes and many more on-field information. It is very difficult to incorporate sophisticated encrypted algorithm in IoT devices due to limited memory, limited processing capability with by consuming less energy. In [45], it is discussed that security is the major obstacles in smart farming. The location-based service which is a mandatory requirement for smart farming is vulnerable to device captured attacks. In device captured attack, the hacker captures the device, analyses the cryptographic algorithm, gets unlimited access to the device storage and collects the information. On the other hand, communication layer is exposed to denial service attack (DoS), jamming attack and man in the middle attack. Further, cloud server is prone to unauthorised services, data tampering, session hijacking and database issues that can hamper cloud service infrastructure. In IoT, there are three mandatory requirements such as authentication, confidentiality and access control. IoT must ensure the aggregation of data in the network layer, authorised access in the application layer and protection of data and devices at perception layer. RFID security issues are usually related to leakage of information, which can reveal the location and other sensitive data. APIs, middleware integration and machine-to-machine communication are some of the terms which are creating the complexity and new risks of security in higher level. Therefore, suitable encryption/decryption algorithms, key distribution mechanisms, intrusion detection systems and secure routing policies must be deployed, considering the limitations of smart hardware devices. IoT connects huge number of smart devices that provide multiple distributed controlling points. These decentralised points are prone to malware which creates a negative impact

on the smart agriculture and becoming the main challenge in IoT-based agricultural sector. Further, less expensive devices are more prone to tampering attack. As IoT devices are distributed by nature and deployed in diverse environments, current security protocols are not enough to provide adequate security. So more encryption and decryption private key algorithms must be designed to provide support to IoT-based real-time applications.

4 Benefits and Pitfalls of IoT-Enabled Agricultural Systems

The agriculture is getting automated day by day on IoT platform. This is reducing the work of farmers and expecting the optimised crop production. Basically, it is very useful for community farming in rural areas [46]. IoT-based agriculture performs well by gathering the information from soil, humid level, and temperature. Monitoring can be done in a regular basis through technological advancement which is helpful in predicting the ecological factors. IoT together with cloud can improve the efficiency of country's agricultural production. From various researches, it is observed that the IoT-enabled agricultural system can be further extended in many ways. Few parameters such as reliability, scalability can be improved. The open-source programming languages like R and Python could be used as a programming language. The development of smart irrigation system could be implemented in other plantations and the data set can be still increased to improve the accuracy of the system. In authentication scheme, the complexities can be reduced further without compromising the security features. The entire work can even be merged with cloud computing environment. From the previous work, some of the new decisions can be made in crops. The data set is maintained for every smart work in agriculture and can be used for further reference.

The country lacks in good agriculture and it could be made still smarter. Using drone with all the weather and temperature information, the type of crop which has to be planted in agriculture can be found. Which crop suits to which environment, that historical information can be found and send to agricultural experts. With those data, the farmer can plant new crops. If the field has the capability to grow by spreading the seeds, it can also be automated. A new device may be invented and made to spread the seeds across the fields based on soil type information. If the climate changes, it can also be intimated through intelligent systems so that some different seeds can be spread. Big data plays a great role in maintaining the data set for weather information, soil type characteristics, etc. Based on the collected data, the seeds can be thrown by agricultural experts or by drone like device to spray the seeds. Another important challenge is that the research has shown that the type of fertiliser can be identified for a typical soil. Despite the huge benefits of IoT-based agriculture, there are lack of awareness among the farmers about the IoT technology. Mostly, farmers are uneducated in rural areas. Young generation is not interested in farming. Our education system does not support in curriculum the intelligence of farming.

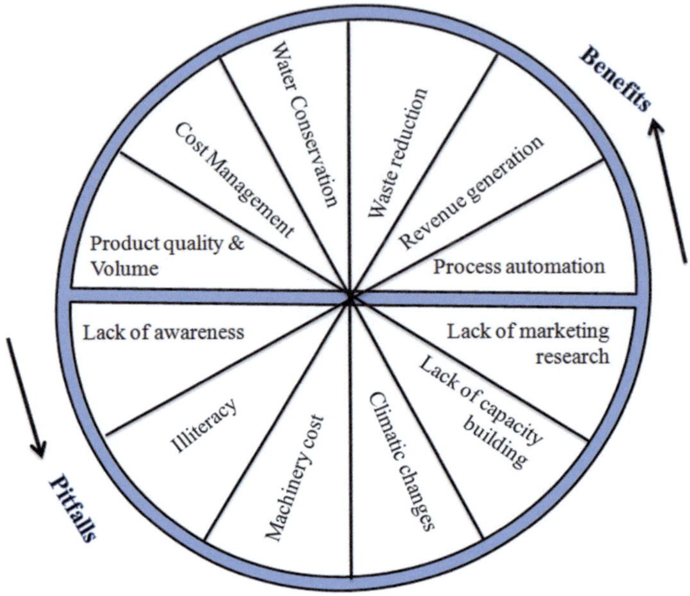

Fig. 9 Benefits and pitfalls of IoT-enabled architecture

Sometimes high-cost machinery may not be affordable by farmers. There is a lack of sale distribution information, lack of marketing research, drastic changes in climatic conditions, more manual work and less knowledge about the weather forecasting may not maximise the productivity (Fig. 9).

5 Conclusion

The interconnected farm called as 'smart farming' is the future of IoT-based agricultural farming. The concept is widespread and creating significant attention in agricultural sector. It seems IoT is changing in agricultural practices in daily basis. Farmers can work their land anywhere taking the real-time data related to any assets from crops, machinery and markets. All can be monitored without the presence of the farmer. The technology today can view data on crops and soils, predict and prevent diseases. Even, the crop can be fed and watered without human presence. Vital data can be stored in the cloud and can be fetched from the cloud at the convenience of the farmers. The sensors are becoming smaller, smarter and cheaper day by day. Network is becoming more secure and intelligent. The future of farming lies in connecting, collecting and analysing the big data to maximise the productivity and efficiency. It is expected that licensed low-power wireless access (LPWA) is the game changer for smart farming due to two reasons. First, due to its well geographical coverage and second is economical. Although unlicensed spectrum is freely available and therefore appealing as a solution, there are significant drawbacks to its use in practice, as its ability to deliver a guaranteed quality of service is compromised by issues of interference and congestion. The regulatory restrictions—which might vary from market to market—will make it difficult for unlicensed spectrum solutions to generate economies of scale. Consequently, narrow band IoT (NB-IoT) has strong industry support as an effective global standard for LPWA connectivity. It has the potential to deliver a step change in smart agriculture by changing the industry's perceptions of what the Internet can provide efficiency connectivity from a wide range of sensors with long battery life over reliable, low-cost, secure, licensed spectrum.

References

1. Xu, L.D., et al.: Internet of Things in industry: a survey. IEEE Trans. Ind. Inf. **10**(4), 2233–2243 (2014)
2. Pujari, J.D., et al.: Image processing-based detection of Fungai diseases in plants. Proc. Comput. Sci. **46**, 1802–1808 (2015). ISSN: 1877-0509
3. Machina: Accessed on 1 Oct 2017 [On-line]. Available: https://machinaresearch.com/reports
4. Tpngke, F.: Smart agriculture based on cloud computing and IoT. J. Conver. Inf. Technol **8**(2) (2013)
5. Ngu, A.H., Gutierrez, M., Metsis, V., Nepal, S., Sheng, Q.Z.: IoT middleware: a survey on issues and enabling technologies. IEEE Int. Things J. **4**(1), 1–20 (2017)
6. Liu, C.H., Yang, B., Liu, T.: Efficient naming, addressing and profile services in Internet of Things sensory envirnoments. Ad Hoc Netw. **18**, 85–101 (2014)
7. Atzori, L., Lera, A., Morabito, G.: The Internet of Things: a survey. Comput. Netw. **54**(15), 2787–2805 (2010)
8. Yan-e, D.: Design of intelligent agriculture management information system based on IoT. In: 2011 Fourth International Conference on Intelligent Computation Technology and Automation, vol. 1, pp. 1045–1049, Mar 2011
9. Wu, Z., Li, S., Yu, M., Wu, J.: The actuality of agriculture Internet of Things for applying and popularizing in China. In: Proceedings of the International Conference on Advances in Mechanical Engineering and Industrial Informatics (EII'15) (2015)
10. Parameswaran, G., Sivaprasath, K.: Arduino based smart drip irrigation system using Internet of Things. Int. J. Eng. Sci. Comput. **6**, 5518–5521 (2016). https://doi.org/10.4010/2016.1348
11. Muhammad, Abubakr, Haider, Bilal, Ahmad, Zahoor: IoT enabled analysis of irrigation rosters in the Indus basin irrigation system. Proc. Eng. **154**, 229–235 (2016)
12. Fang, S., Da Xu, L., Zhu, Y., Ahati, J., Pei, H., Yan, J., Liu, Z.: An integrated system for regional environmental monitoring and management based on Internet of Things. IEEE Trans. Ind. Inf. **10**(2), 1596–1605 (2014)
13. Dursun, M., Ozden, S.A.: A wireless application of drip irrigation automation supported by soil moisture sensors. Sci. Res. Essays **6**, 1573–1582 (2011)
14. Khattab, A., Abdelgawad, A., Yelmarthi, K.: Design and implementation of a cloud-based IoT scheme for precision agriculture. In: 28th International Conference on Microelectronics, pp. 201–204 (2016)
15. Kodali, R.K., Sahu, A.: An IoT based weather information prototype using WeMos. In: 2nd International Conference on Contemporary Computing and Informatics, pp. 612–616 (2013)
16. Bing, F.: The research of IOT of agriculture based on three layers architecture. In: 2nd International Conference on Cloud Computing and Internet of Things, pp. 162–165 (2016)
17. Edwards-Murphy, F., Magno, M., Whelan, P.M., O'Halloran, J., Popovici, E.M.: B+WSN: smart beehive with preliminary decision tree analysis for agriculture and honey bee health

monitoring. Comput. Electron. Agric. **124**, 211–219 (2016). https://doi.org/10.1016/j.compag.2016.04.008
18. Ding, Q., Ma, D., Li, D., Zhao, L.: Design and implementation of a sensors node oriented water quality monitoring in aquaculture. Sens. Lett. **8**(1), 70–74 (2010)
19. Bang, J., Lee, I., Noh, M., Lim, J., Oh, H.: Design and implementation of a smart control system for poultry Breeding's optimal LED environment. Int. J. Control Autom. **7**(2), 99–108 (2014). https://doi.org/10.14257/ijca.2014.7.2.10
20. Vernandhes, W., Salahuddin, N.S., Kowanda, A., Sari, S.P.: Smart aquaponic with monitoring and control system based on IoT. In: 2017 Second International Conference on Informatics and Computing (ICIC), pp. 1–6. IEEE (2017)
21. Brewster, C., et al.: IoT in agriculture: designing a Europe-wide large-scale pilot. IEEE Commun. Mag. **55**(9), 26–33 (2017)
22. CLAAS: Accessed on 20 Sept 2017 [Online]. Available: http://www.claasofamerica.com/
23. Yang, K., Liu, H., Wang, P., Meng, Z., Chen, J.: Convolutional neural network-based automatic image recognition for agricultural machinery. Int. J. Agric. Biol. Eng. **11**(4), 200–206 (2018)
24. Karim, F., Karim, F.: Monitoring system using web of things in precision agriculture. Proc. Comput. Sci. **110**, 402–409 (2017)
25. Johannes, A., et al.: Automatic plant disease diagnosis using mobile capture devices, applied on a wheat use case. Comput. Electron. Agric. **138**, 200–209 (2017). ISSN: 0168-1699. https://doi.org/10.1016/j.compag.2017.04.013. http://dx.doi.org/10.1016/j.compag.2017.04.013
26. Petrellis, N.: A smart phone image processing application for plant disease diagnosis. In: 2017 6th International Conference on Modern Circuits and Systems Technologies (MOCAST), pp. 1–4 (2017). ISSN: 1613-0073. https://doi.org/10.1109/mocast.2017.7937683. http://ieeexplore.ieee.org/document/7937683/
27. Kawakami, Y., et al.: Rice cultivation support system equipped with water-level sensor system. IFAC-PapersOnLine 49(16), 143–148 (2016). ISSN: 2405-8963. https://doi.org/10.1016/j.ifacol.2016.10.027. http://dx.doi.org/10.1016/j.ifacol.2016.10.027
28. Sarangi, S., Umadikar, J., Kar, S.: Automation of agriculture support systems using Wisekar: case study of a crop-disease advisory service. Comput. Electron. Agric. **122**, 200–210 (2016)
29. Rodriguez, S., Gualotuna, T., Grilo, C.: A system for the monitoring and predicting of data in precision agriculture in a rose greenhouse based on wireless sensor networks. Proc. Comput. Sci. **121**, 306–313 (2017)
30. Dan, L., et al.: Intelligent agriculture greenhouse environment monitoring system based on IOT technology. In: 2015 International Conference on Intelligent Transportation, Big Data and Smart City (ICITBS). IEEE (2015)
31. Akkaş, M.A., Sokullu, R.: An IoT-based greenhouse monitoring system with Micaz motes. Proc. Comput. Sci. **113**, 603–608 (2017)
32. Jagdale, T., Mali, M.B.: Greenhouse wireless network monitoring and management using IoT. Int. J. Innov. Res. Electr. Electr. Instrum. Control Eng. (2016)
33. Birla, A., Biral, A., Centenaro, M., Zanella, A., Vangelista, L., Zorzi, M.: The challenges of M2M massive access in wireless cellular networks. Digital Commun. Netw. **1**(1), 1–19 (2015)
34. IERC: European Research Cluster on the Internet of Things. Technical report [Online]. Available at http://www.internet-of-things-research.eu/pdf/IERC_position_paper_IoT_semantic_interoperability_final.pdf
35. Tayur, V.M., Suchitra, R.: Review of interoperability approaches in application layer of Internet of Things. In: 2017 International Conference on innovation Mechanisms for Industry Applications (ICIMIA), pp. 322–326, Feb 2017
36. Open connectivity: Accessed on 20 Sept 2017 [On-line]. Available: https://openconnectivity.org/
37. IFTTT: Accessed on 20 Sept 2017 [Online]. Available: https://iftt.com/
38. Zhao, J.C., Zhang, J.F., Feng, Y., Guo, J.X.: The study and application of the IoT technology in agriculture. In: 2010 3rd International Conference on Computer Science and Information Technology, vol. 2, pp. 462–465, July 2010

39. Shinde, T.A., Prasad, J.R.: IoT based animal health monitoring with naive bayes classification. IJETT 1(2) (2017)
40. Godfray, H.C.J., Beddington, J.R., Crute, I.R., Haddad, L., Lawrence, D., Muir, J.F., Pretty, J., Robinson, S., Thomas, S.M., Toulmin, C.: Food security: the challenge of feeding 9 billion people. Science **327**(5967), 812–818 (2010)
41. Nelleman, C., et al.: The environmental food crisis. In: The Environment's Role in Averting Future Food Crises. A UNEP Rapid Response Assessment. United Nations Environment Program, GRID-Arendal, Arendal, Norway (2009)
42. Deepak, V., Megha, T., Prithvi, G.H., Syed, S.A., Sharavana, K.: Cold storage management system for farmers based on IoT. Int. J. Recent Trends Eng. Res. **3**(5), 556–561 (2017)
43. Centenaro, M., et al.: Long range communications in unlicensed bands: the rising stars in the IoT and smart city scenarios. IEEE Wireless Commun. **23**(5), 60–67 (2016)
44. Adhikary, A., Lin, X., Wang, Y.P.E.: Performance evaluation of NB-IoT coverage. In: 2016 IEEE 8th Vehicular Technology Conference (VTC-Fall), pp. 1–5, Sept 2016
45. Sicari, S., Rizzardi, A., Grieco, L.A., Coen-Porisini, A.: Security, privacy and trust in Internet of Things: the road ahead. Comput. Netw. **76**, 146–164 (2015)
46. Bo, Y., Wang, H.: The application of cloud computing and the Internet of Things in agriculture and forestry. In: 2011 International Joint Conference on Service Sciences, May 2011, pp. 168–172

Land Suitability Analysis for Peri-urban Agriculture Using Multi-criteria Decision Analysis Model and Crop Condition Monitoring Methods: A Case Study of Kolkata Metropolitan Area

Sushobhan Majumdar

Abstract In recent decades urban and peri-urban agriculture have gained attention worldwide as these are the store house of major cities for the availability of various resources like milk, fish, flower, etc. Urban or peri-urban farming contributes to food security and income generation for the people who resides around the major city areas. Often it faces various problems like encroachment of urban area, polluted water, lack of efficient workers, etc., especially in the developing countries. In this paper, multi-criteria decision analysis model and crop condition monitoring method have been applied to find out the suitable zones of peri-urban agriculture around Kolkata city. This model was tested on the Kolkata Metropolitan Area (KMA) using various criteria like land use, digital elevation model (DEM), water facility, road and market facility, etc. For analyzing the land use of KMA both supervised classification methods and to find out the agricultural area crop condition monitoring method (e.g., NDVI, i.e., normalized difference vegetation index) has been used by using remote sensing images. For the analysis of water facility, road and market facility in the various areas of KMA census data has been used. Multi-criteria decision analysis model revealed that northwest, central east, southeast, and northern zones of Kolkata Metropolitan Area is the most suitable zones for peri-urban agriculture. Finally, it can be said that this model is able to allocate the suitable land for the peri-urban agriculture very precisely. This model will help the urban, peri-urban planners, policy-makers, and decision-makers for taking action on various decisions at different levels.

1 Introduction

Urbanization and population growth are mainly responsible for the land use and land cover change of an area [19]. Urbanization and urban growth concentrate mainly on the peripheral areas, rural areas, and small town areas in the outskirt of the city [1]. Population growths in urban areas are related to the spatial expansion of cities [14,

S. Majumdar (✉)
Department of Geography, Jadavpur University, Kolkata, India
e-mail: sushobhan91@gmail.com

P. K. Pattnaik et al. (eds.), *IoT and Analytics for Agriculture*, Studies in Big Data 63,
https://doi.org/10.1007/978-981-13-9177-4_8

17]. Due to huge urbanization in the metropolitan city area and its resultant urban expansion, most of the vegetation area, agricultural land, and water bodies have been converted into the residential area, commercial area, or barren land area [20]. It effects on the natural ecosystems of the local environments of the city. Peri-urban areas are basically the transition zone between the rural and urban areas as there are no specific boundaries between these. According to Guerois and Pumain [8], this area is due to high costs of flat, need of individual dwellings and high land rent. This zone is an area of unplanned growth and urban areas have lost resident population gaining population to transit [3]. One of the striking features of this area is people used to commute to city from these areas [6]. Another main feature of this area is low density of urbanization [4].

Rapid increase in the limit of urban agglomerations have raised the concern to monitor the urban and peri-urban landscapes in the context of sustainable land use planning and management [13], environmental and developmental issues [7, 10, 15]. Most of the studies on the metropolitan areas have either focused either on the urban areas or in the rural areas. But in recent years, urban and peri-urban agriculture have gained attention in worldwide phenomenon. Urban farming provides food security and income generation for the residents of both urban and rural areas and it also protects the city from pollution and different types of city externalities. But in the developing countries like India, peri-urban agriculture faces many problems like polluted water, low supply of irrigated water, less uses of insecticides, lack of modern technology, low education of the farmers, etc.

In this paper, an attempt has been made to find out the suitable zones for peri-urban agriculture using multi-criteria decision analysis model and crop condition monitoring method. The objectives are to (1) characterize the agricultural area of Kolkata Metropolitan Area (KMA) using crop condition monitoring method, (2) find out the suitable areas for agriculture in KMA and reasons behind them.

2 Study Area

Kolkata city and its surroundings area are named as Kolkata Metropolitan Area. All types of planning and development of this area are under the jurisdiction of Kolkata Metropolitan Development Authority (KMDA). For study purpose Kolkata Metropolitan Area (KMA) has been considered. It is one of the largest agglomerations in the eastern India and also one of the oldest. KMA extends from Kalyani to Budge Budge in the eastern bank and Bansberia to Uluberia in the western bank of the River Hooghly. It covers six districts in West Bengal, i.e., Kolkata, Howrah, Hooghly, North 24 Parganas, South 24 Parganas, and Nadia. According to the Census 2011, total population of KMA is 1 crore and 58 lakhs comprising an area of 1841.47 km^2. Between the year 2001 and 2011, growth rate population is 10.30% (Table 1; Fig. 1).

Table 1 Present composition of KMA

Units	No's of units	Area (km^2)
Municipal corporations	3	272.32
Municipalities	39	667.28
Census towns	155	364.53
Out growth's	6	4.68
Rural areas	394	532.66
Total KMA		1841.47

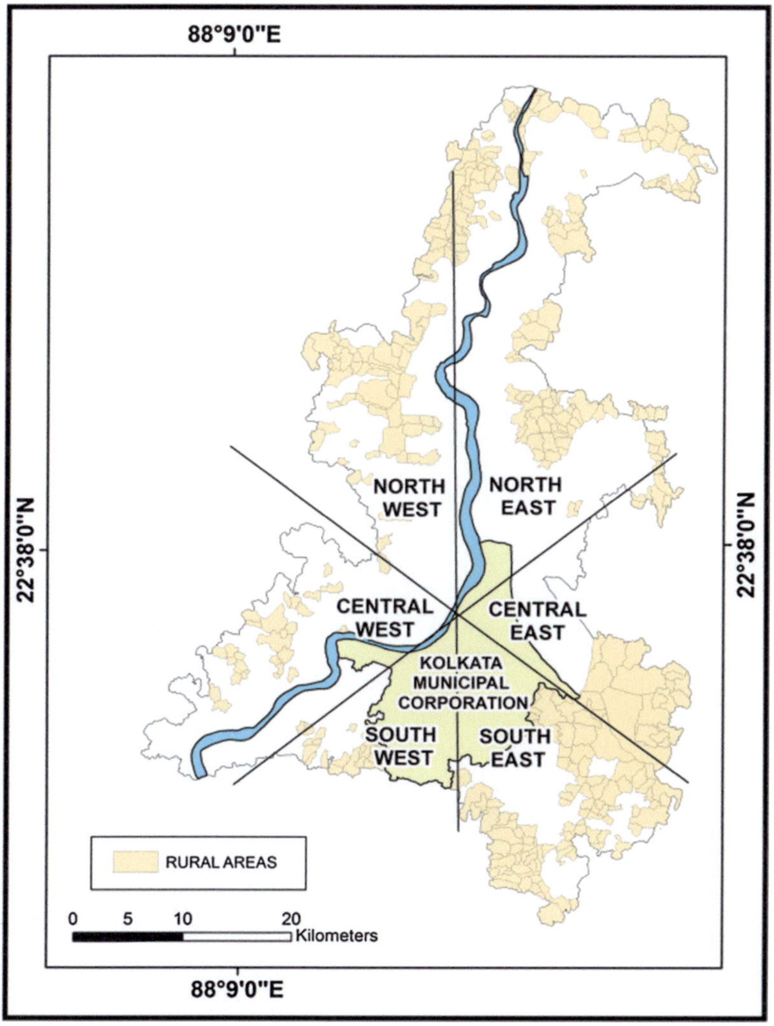

Fig. 1 Map of Kolkata Metropolitan Area

3 Database and Methodology

The data set for this study is comprised of four Landsat images recorded from 1990 to 2015. Detailed description of those images discussed below (Table 2).

Four sets of landsat satellite images were used here. Two sets of Landsat Thematic Mapper (TM) and Enhanced Thematic Mapper (ETM+) images, and OLI 8 image (with path/row 138/45) have been used. At the time of layer stacking of those images thermal band has been excluded. Digital maps published from the Kolkata Metropolitan Authority have been digitized and geo-referenced from hardcopy topographic maps with the scale of 1:50,000. This map has been used as a reference image. It is mainly used for geometric correction of those satellite images and to collect some ground truth information. Finally, ground information was collected between 1990 until 2015 to get land use/land cover information properly. Then supervised classification algorithm was used to scrutinize the land cover types.

Image Processing

Those remotely sensed satellite images were geometrically corrected and geo-coded to the Universal Transverse Mercator (UTM) coordinate system projection by using a reference image. Erdas imagine 9.2 software which have been geo-referenced previously by the topographical sheets provided by the Survey of India (SOI). Minimum of 450 randomly distributed ground control points (GCPs) were selected from the topographical sheets for geo-referencing the image accurately. Re-sampling technique was performed using a nearest neighbor algorithm technique. Image transformation technique was used with root mean square (RMS) error of 0.1 pixels indicating that the image was accurate to within one pixel.

Supervised Classification

Land use and land cover classes are mapped from the satellite images through the process of a supervised classification [5, 18]. But major objective of the image classification procedure is to automatically classify all pixels in an image into land cover classes or themes [12].

For digital image processing and for the supervised classification of the Landsat images, Erdas imagine 9.2 software has been used. Training samples were selected for each of the predetermined LULC types by feature space tool around the representative areas with the help of feature space tool. A spectral signature is considered to be

Table 2 Detailed information of utilized satellite imagery

Satellite	Acquisition date	Sensor	Spatial resolution (m)	Projection
Landsat 8	08-03-2015	OLI-TIRS	30	WGS 84 UTM 45 N
Landsat 7	21-01-2010	TM	30	WGS 84 UTM 45 N
Landsat 7	17-11-2000	ETM+	30	WGS 84 UTM 45 N
Landsat 5	14-11-1990	TM	30	WGS 84 UTM 45 N

Source USGS.gov.in

satisfactory when confusion among the land covers to be mapped is minimal [9]. After collecting those spectral signatures from each of the representative land cover classes image classification have been done using maximum likelihood algorithm as a classification method.

Maximum livelihood algorithm is one of the common parametric classifiers used for especially in case of supervised classification. This algorithm is used for computing the weighted distance or likelihood (D) of unknown measurement vector (X) belonging to one of the known classes (M_c) which is based on the Bayesian equation [16].

$$D = \text{In}\ (a_c) - [0.5\ \text{In}(|\text{cov}_c|)] - \ [0.5(X - M_c)\,T(\text{cov}_e - 1)(X - M_e)]$$

After the field verification only five classes have been identified as these five are the major land use and land cover features in this area which has been categorized into various subcategories also (Table 3).

To scrutinize the crop condition in the peripheral areas of the city normalized difference vegetation index (NDVI) has been used [2, 11] (Bechtal et al. 1997; El-Shikha et al. 2007).

$$\text{NDVI} = \text{NIR} - \text{RED/NIR} + \text{RED}$$

where “NIR” band represents near-infrared band and “RED” band represents red band of the Landsat images. NDVI profile depends on the pixel value of the processed image, which reflects the changes of the crop including planting, seeding, harvesting, etc., of the processed images. Crop condition with the cultivating area can be easily calculated by NDVI profile using spectral signatures of the vegetation area. To scrutinize the slope of the total area digital elevation model (DEM) and contour map have been used using satellite images.

To find out the suitable places for the agriculture different criteria’s have been taken like primary workers, percentage of agricultural workers, percentage of vegetation, piped water supply, availability of drainage facility, having market facility, availability of bus services, and distance from the railway (within 5 km) using Census 2011 data among the rural areas of KMA. Values for different categories have been selected after averaging the value of both KMA and West Bengal. In KMA, the

Table 3 Identified classes by supervised classification

No	Land use classes	Description
1	Built-up area	Residential, commercial, industrial, roads, railway, mixed urban or built-up area
2	Vegetation	Vegetative areas and forest area
3	Agricultural land	Agricultural areas and cultivated land
3	Water body	River, pond, canal, reservoir
4	Barren land	Barren land and fallow land

area other than the urban area has been considered as the areas of the study. To find out suitable places for agriculture within the KMA "land suitability index" has been done using standardized scoring method (Table 4).

4 Discussion and Analysis

In the rural areas of the Kolkata Metropolitan Area, it has been found that percentage of cultivated land is high in the peripheral areas of the city. Applying NDVI techniques for the crop condition monitoring methods it have been found that major crops of these areas are rice, potato, jute, etc. Crop condition among the crops has been calculated using NDVI techniques.

Applying NDVI techniques it have been found that in 1990 (Fig. 2), maximum values for NDVI is 0.72 and minimum is −0.91. From the NDVI images of 1990, it has been found that percentage of agricultural land is high in the northeastern, northwestern, southeastern, and southwestern part of the Kolkata city.

In 2000 (Fig. 2), in the NDVI images of KMA, maximum values for NDVI was 0.61 and minimum was −0.61. From this indices image, it has been found that the percentage of cropping land is high in the northeast, northwest, and southeast part of the Kolkata city.

In 2010 (Fig. 2), values for the NDVI ranges from 0.66 to −0.39. In this time period, percentage of cropland is high near the Kalyani, Gayespur area in the north-eastern part of KMA, toward the Bansberia and rural areas of Hooghly-Chinsurah

Table 4 Land suitability index scoring pattern

Factors	Criteria	Score	Factor	Elevation	Score
Percentage of cropland to total cultivated area, percentage of primary workers to total primary workers, percentage of agricultural area to total cultivated area, percentage of vegetation to total vegetation, average water supply and drainage facility, average market facility, average bus service and railway facility	Low/bad	1	Elevation	Low	5
	Moderately low/bad	2		Moderately low	4
	Medium	3		Medium	3
	Moderately high/good	4		Moderately high	2
	High/good	5		High	1

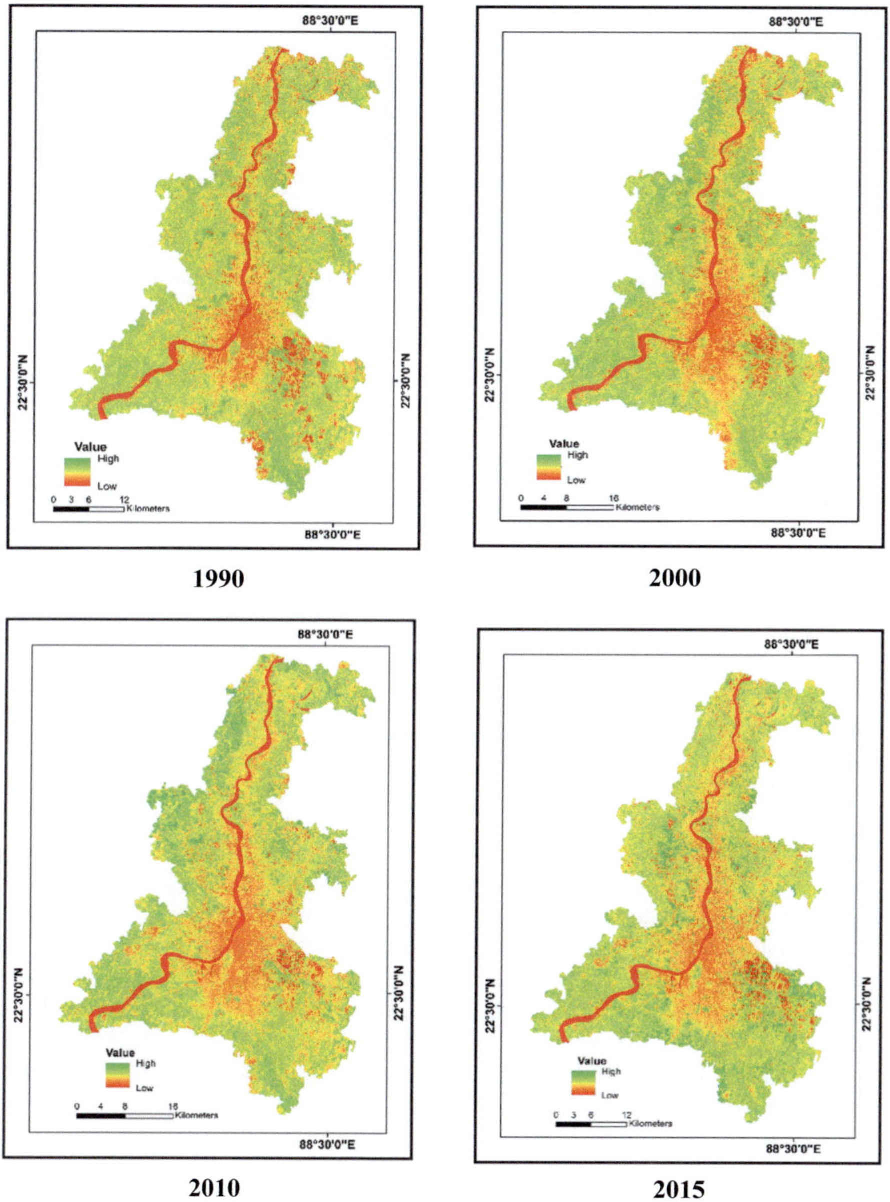

Fig. 2 Crop Condition Monitoring Method: Normalized Difference Vegetation Index (NDVI) of the four time periods

block, in the central western part of the Kolkata city, i.e., in the rural areas of the Howrah district.

In 2015 (Fig. 2), in the NDVI images of KMA, maximum value is 0.54 and minimum value is −0.21. From the NDVI values of KMA, it has been found that percentage of agricultural areas or crop areas is high in the northeastern part (i.e., near the census towns like Jethia, Chendua beside the Barrackpore-Kalyani Expressway) of the KMA, toward the northwestern part of the Kolkata (i.e., beside the Polba Dadpur area beside the Delhi road), and in the southeastern part of the Kolkata city (i.e., in the rural blocks of Baruipur and Sonarpur blocks of South 24 Parganas). But percentage of cultivated land or agricultural land has been decreased very rapidly because of the rapid encroachment of the urban areas toward the rural areas of the Kolkata city (Fig. 3).

From the digital elevation model (DEM) and contour map (Fig. 4) of the whole KMA, it has been found that slope and elevations of the entire area is very nearly uniform as it is located in the deltas of Hooghly River. Some areas of the KMA like northwest, southeast area are low lying area which is suitable for the agricultural purposes. Other than the slope factor different factors like easy accessibility from the other areas both from the Kolkata and district headquarter and nearness of Kolkata city are the major factors behind the development of these agricultural areas.

From the land use and land cover map of Kolkata Metropolitan area, it has been found that the percentage of agricultural area and vegetation is much high in the peripheral areas of the city. It decreases toward the city core or nearer to the urban areas. Table 5 showing land use of KMA from 1990 to 2015 (Fig. 3).

Land Use Pattern in 1990

A total of five land use categories has been identified in 1990 image classification those are built-up area, vegetation area, water bodies, cultivated land, and barren land. From this classification, it has been found that vegetation area accounted for 52.25% of the total land use. It is followed by built-up area which comprises 16% of the total area. Barren land, water bodies, and cultivated land covers 14, 9, and 7% of the total area, respectively. In this time period, percentage of agricultural area is high in the northwest, northeast, southeast, southwest, central east and central west part of the KMA (Fig. 3). These zones act as a storehouse of Kolkata city as from these areas large amounts of fruits, fresh vegetables, fishes, milk, and flower used to come to the Kolkata city for the daily need of the city people.

Land Use Pattern in 2000

In 2000, built-up area covered only 26% of the total area. Vegetation area accounted for 39% of the total area. 6% of the area is under the cultivated land, water bodies occupy 5%, and barren land occupies 24% of the total area, respectively (Fig. 3). In this time period, percentage of agricultural area or cultivated land is high nearer the Kalyani, Gayespur area, toward the Barasat area, Sonarpur-Bartuipur area, Dankuni area, Hooghly-Chinsurah area, Uluberia area, Budge Budge-Pujali area, etc.

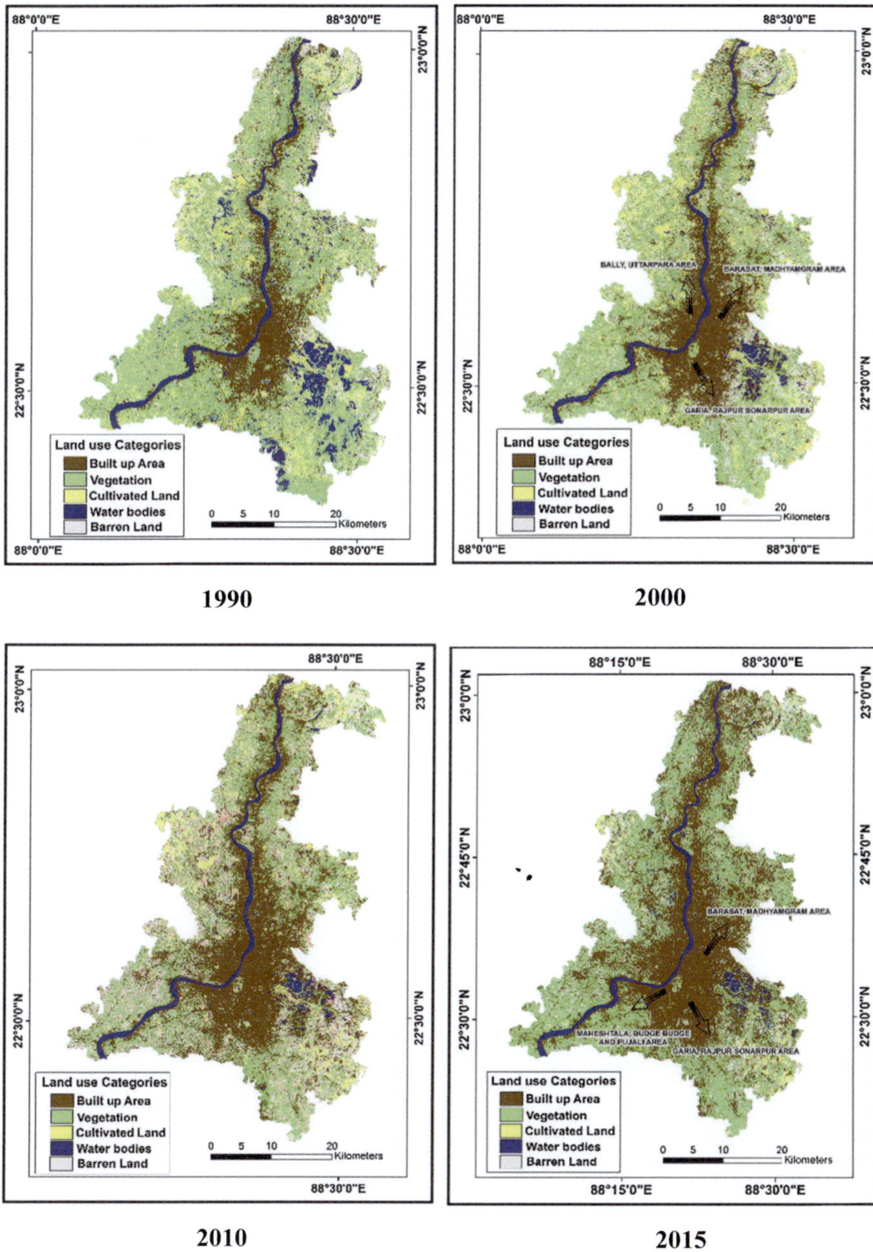

Fig. 3 Land use and land cover pattern of Kolkata Metropolitan Area

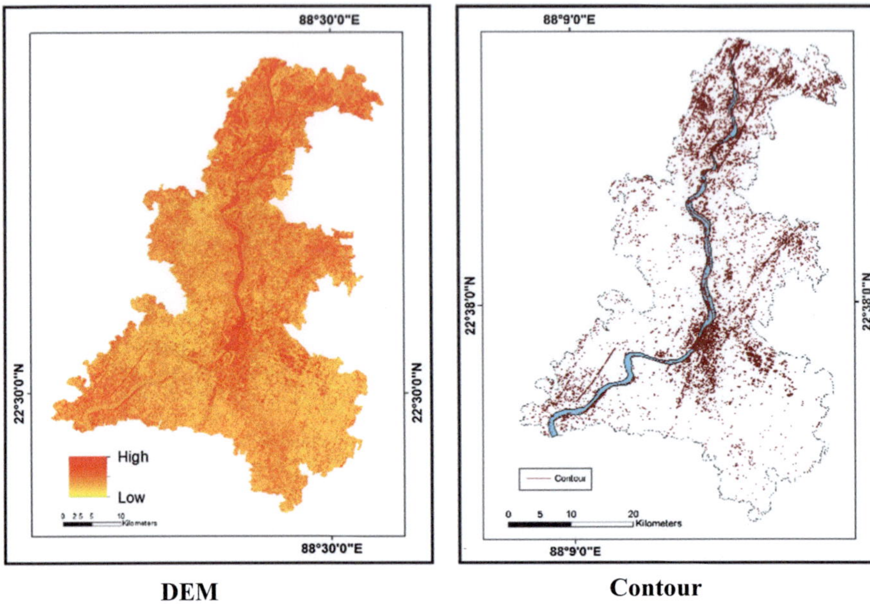

Fig. 4 Digital Elevation Model (DEM) and Contour map showing elevation of KMA

Table 5 Trends of land use and land cover change (percent) in Kolkata Metropolitan Area

Year	1990	2000	2010	2015
Built-up area	16.55	26.04	30.17	33.60
Vegetation	52.25	39.40	38.17	37.11
Cultivated land	7.89	5.58	3.89	3.26
Water bodies	9.32	5.26	4.78	3.80
Barren land	13.90	23.70	23.07	22.23

Source Computed after the classification of satellite images

Land Use Pattern in 2010

In 2010 (Fig. 3), vegetation area covers 38% of the total area. Built-up area accounted for 30% of the total area. Five percent of the land is under water bodies, 4% is under cultivated land, and 23% of the land is under barren land.

Land Use Pattern in 2015

In 2015, 34% of the area is under built up. Thirty-four percent of the area is under the vegetation area.

After the five years, i.e., in the year 2015 (Fig. 3), 34% of the total area was under the built-up area and it is followed by vegetation area (i.e., 37% of the total area), barren land (i.e., 22% of the total area), water bodies (i.e., 4% of the total area), and cultivated land (i.e., 3% of the total area). Between this time periods the

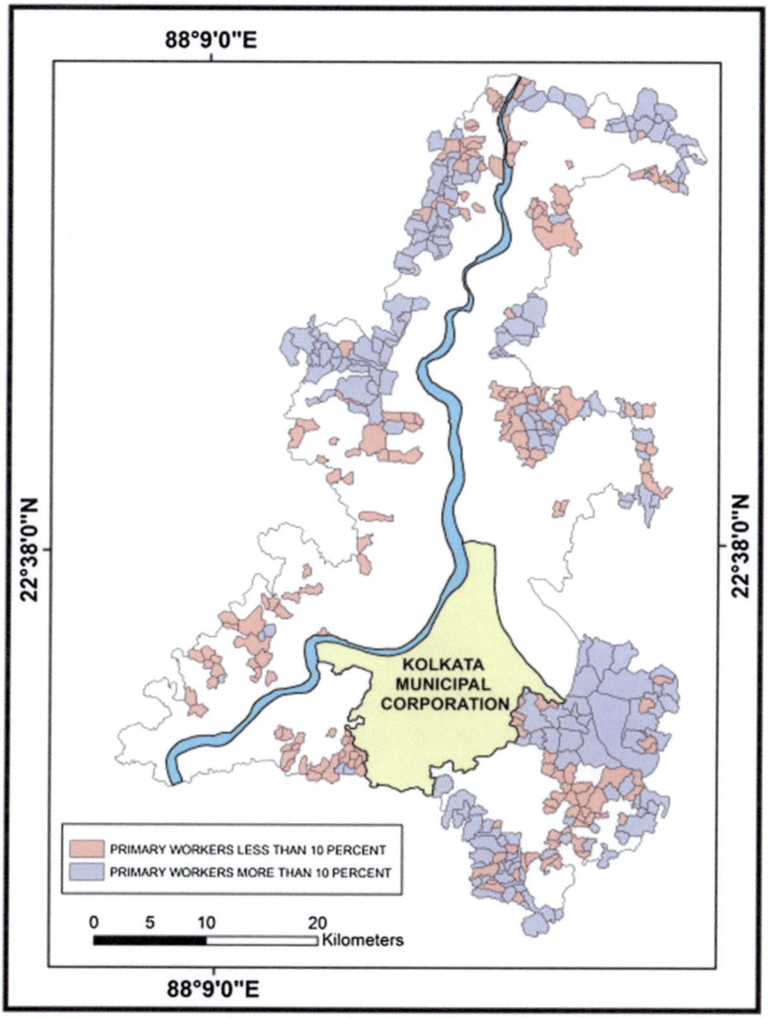

Fig. 5 Percentage of primary workers

expansion has occurred toward the northeastern side, southeastern and especially in the southwestern side (i.e., Maheshtala, Budge Budge, and Pujali) from the city. Due to the expansion vegetation area has also reduced by 19 km^2 between 2010 and 2015. These type of conversion (i.e., from agriculture to built-up area) is much high in Barasat, Bansberia, and Gayespur, etc., municipal area as it has seen from the classification of satellite images.

From the percentage of primary workers (Fig. 5), it have been found that percentage of primary workers is high in the southeastern part of KMA (i.e., in the Sonarpur block of South 24 Parganas block), in the northwestern part of KMA (i.e., in the rural areas of Hooghly district), and some parts of Nadia district and Howrah district. Percentages of primary workers are low in the central eastern and central western part of the Kolkata. From the overview of the primary workers of the entire KMA, it has

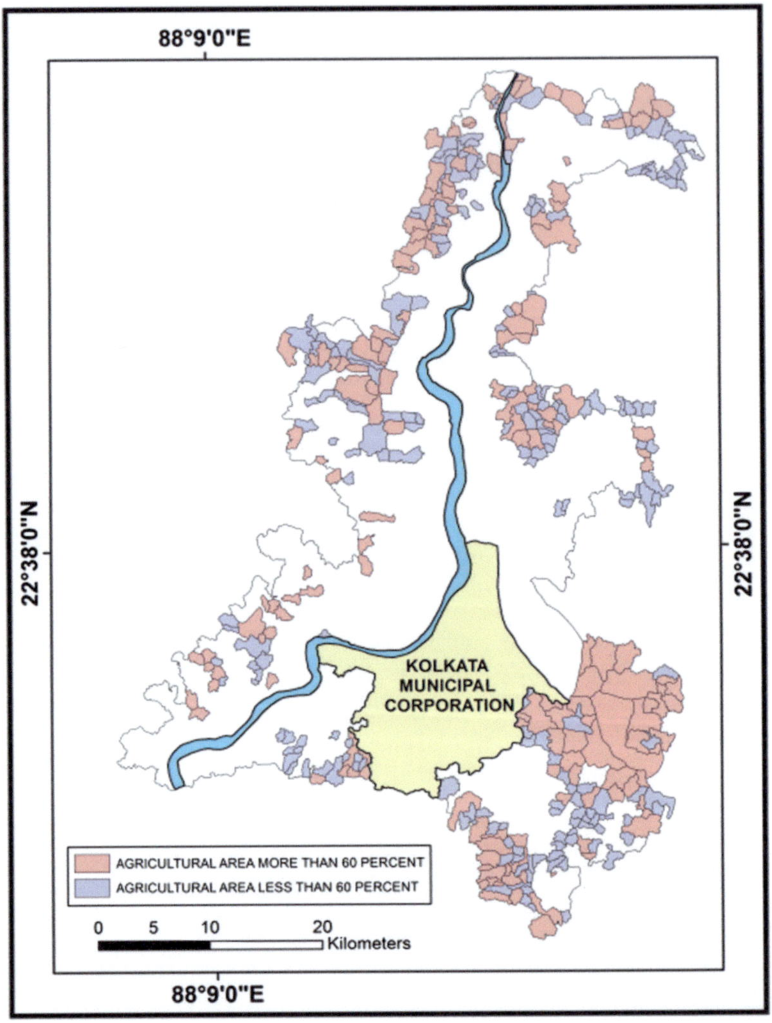

Fig. 6 Percentage of agricultural area

been found that percentages of primary workers are low toward the Kolkata which increases outward. Because of the different reasons like availability of various job opportunities and easy earning of cash money most of the people wants to engage in the secondary activities rather than the primary activities. So percentages of primary workers have been decreased in recent decades (i.e., 2001–2011).

After visualizing the percentage of agricultural area (Fig. 6), it has been found that percentage of agricultural land or cultivated land is high in the southeastern part and northwestern part of KMA. Because of the easy accessibility from the other areas both by the roadway and railway, nearest market facility, fertile soil with low slope, etc., are the major reasons behind the development of the agricultural area. Recently, percentage of agricultural area is decreasing because of the encroachments of the

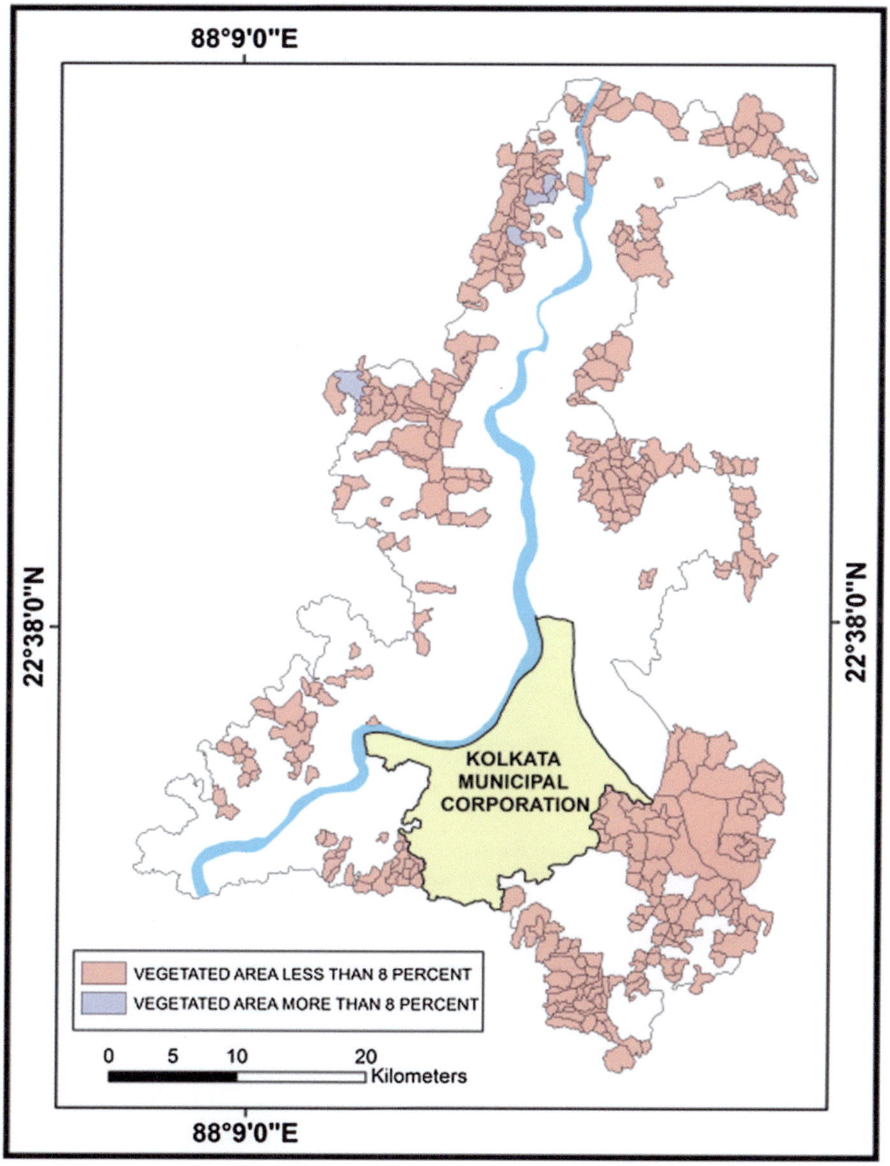

Fig. 7 Percentage of vegetation area

urban areas to rural areas. Other than these reasons, because of the easy availability of cash money, most of the agricultural labor or cultivators in these areas want to engage in the secondary activities like grocery shops, car driving, etc.

From the vegetation percentage (Fig. 7) in entire KMA, it has been found that percentage of vegetation area is very low and it is almost lower than the overall

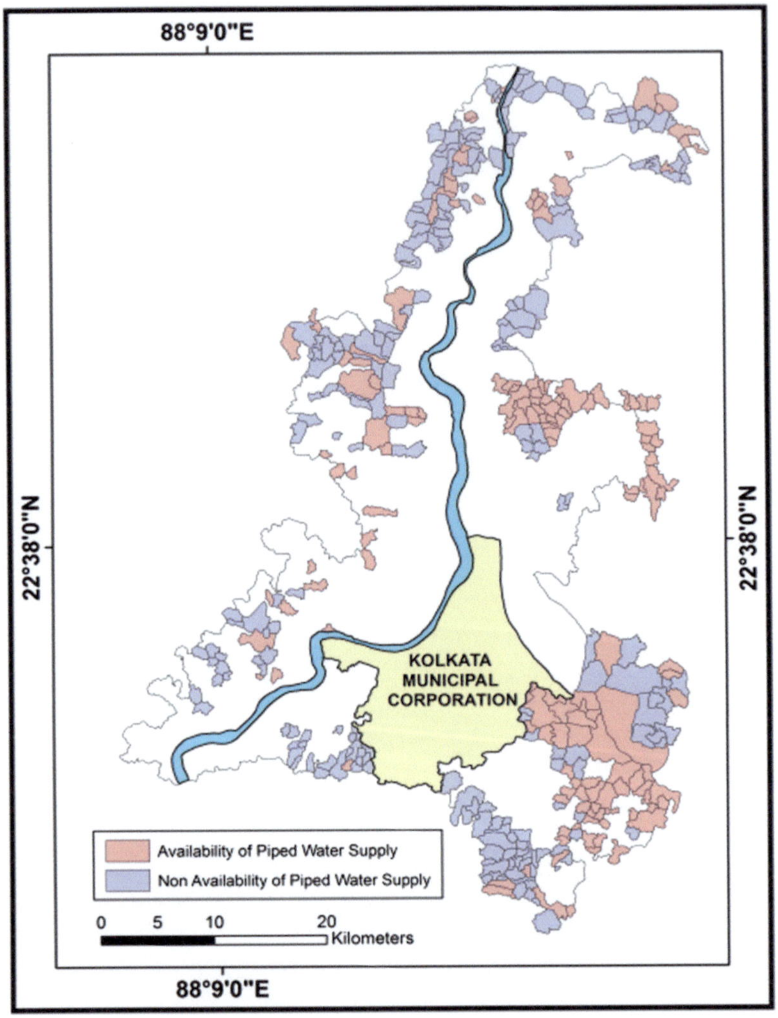

Fig. 8 Piped water supply facility

percentage of West Bengal. It is low because of the expansion of the urban area most of the rural agricultural area or vegetation area has been converted into the residential area. Low percentage of vegetation creates extreme pressure on the local land ecosystem of the area.

By analyzing the piped water supply (Fig. 8) among the rural areas of KMA, it has been found that percentage of non availability of water supply is high. Most of the area nearest to the Kolkata has facility of piped water supply. Piped water supply is very useful for the cultivation of crops. Otherwise cultivators of these areas have to depend on the rain, inland pond, and ground water for their cultivation. Because of the lack of piped water supply facility most of the areas are lacking of good drainage

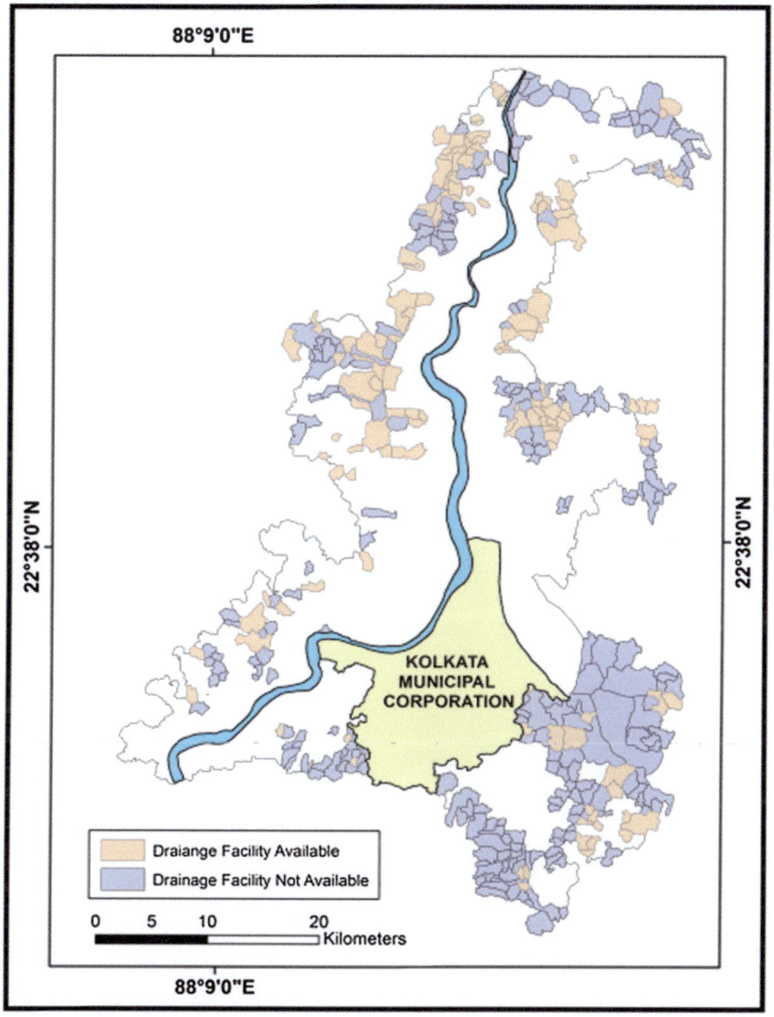

Fig. 9 Availability of drainage facility

facility. Lack of drainage facility (Fig. 9) leads to water logging and water borne diseases in the rural areas.

Good market facility (Fig. 10) is another criterion for the agricultural practices. It not only helps them to earn easy earnings of money but also those goods can export to the city for the daily needs of the people. Market facility is major factor behind the growth of the agricultural area. Some areas in the central eastern part (like Barsat and Madhyamgram), in the northwestern part of KMA (like rural areas of Hooghly-Chinsurah block) have good market facility. From these type of rural

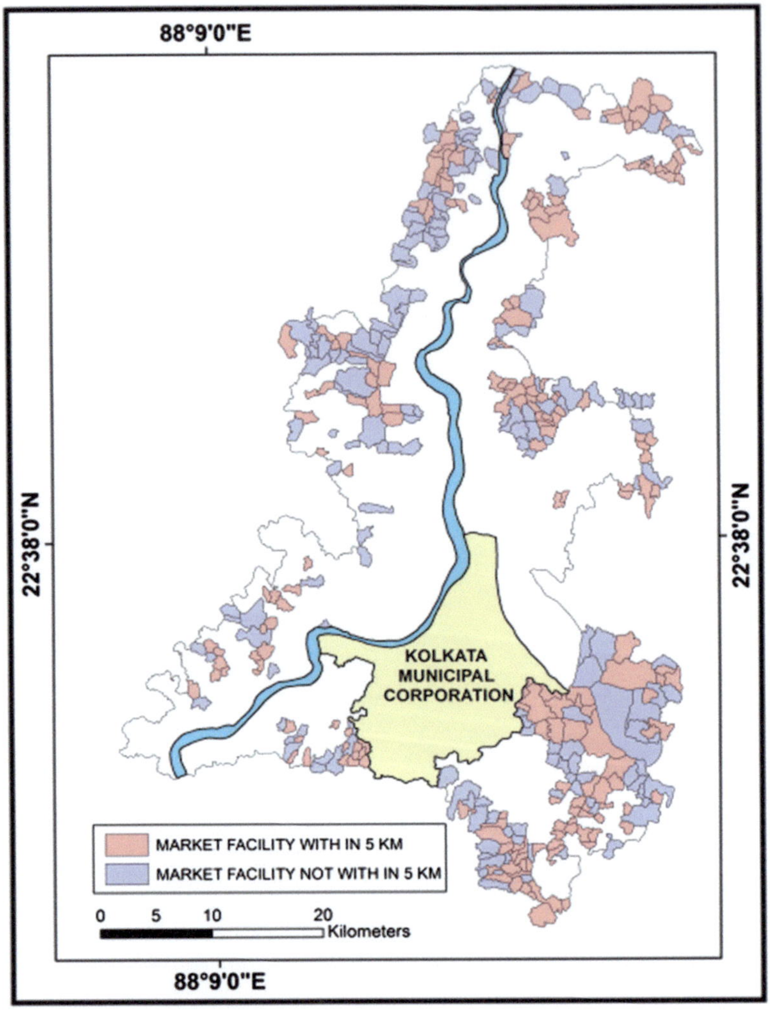

Fig. 10 Availability of market facility

markets cultivators can easily export their goods for the daily needs of the people residing in urban areas.

Availability of bus service is another major criterion behind the growth of agricultural areas, as most of the cultivators in the rural areas of KMA used to transport their crops by the roof of the buses (Fig. 11). From the overview of the availability of bus services facilities in the rural areas of KMA, it has been found that except some of the villages most of the villages have good bus services facility. Except railway facility it is one of the major modes for the transportation of the crops.

From the maps of distances from the railway lines of KMA, it has been found that most of rural mouzas in the Baruipur block of South 24 Parganas district, toward the

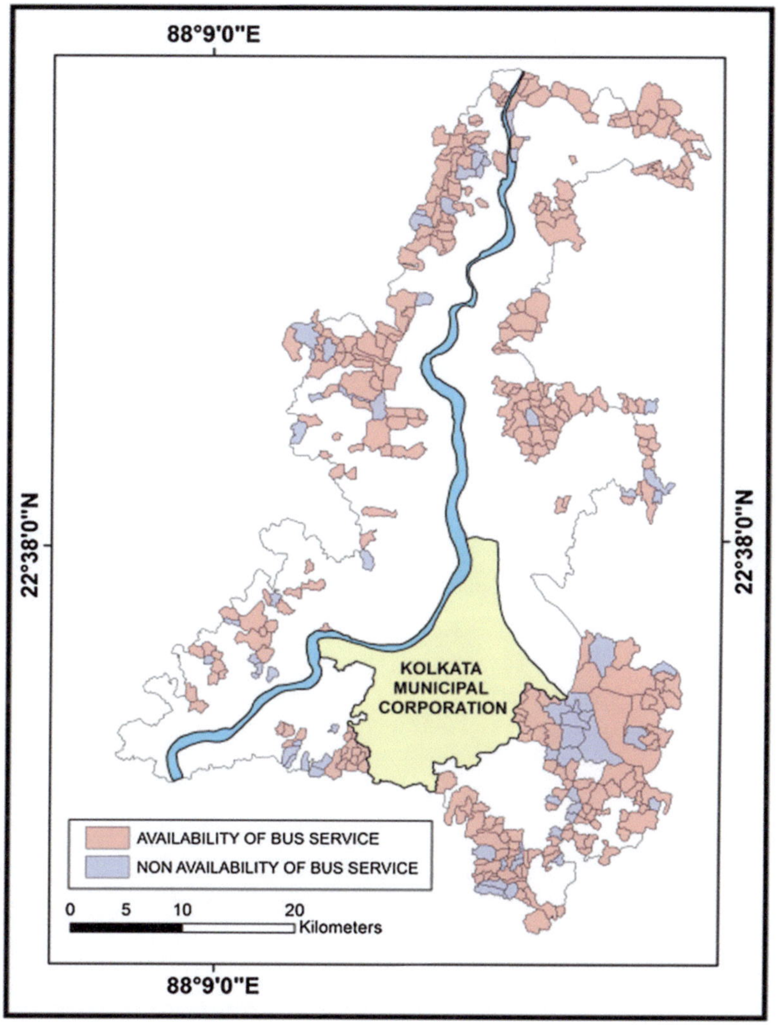

Fig. 11 Availability of bus services

rural areas of the Howrah district the distance from the railway line is within the 5 km distance (Fig. 12). Because of the low transportation costs with less time consuming most of the cultivators prefer this mode of transport especially for the perishable products. Most of the farmers used to export their products by the suburban railway facilities of Kolkata city. These agricultural zones around the Kolkata city act as a green belt of the city. Different products like milk, crops, and fishes used to export to Kolkata and these areas are called as the storehouse of the Kolkata city.

Table 6 shows land suitability analysis using multi-criteria analysis model of KMA. From the land suitability table, it can be seen that in the northeastern part

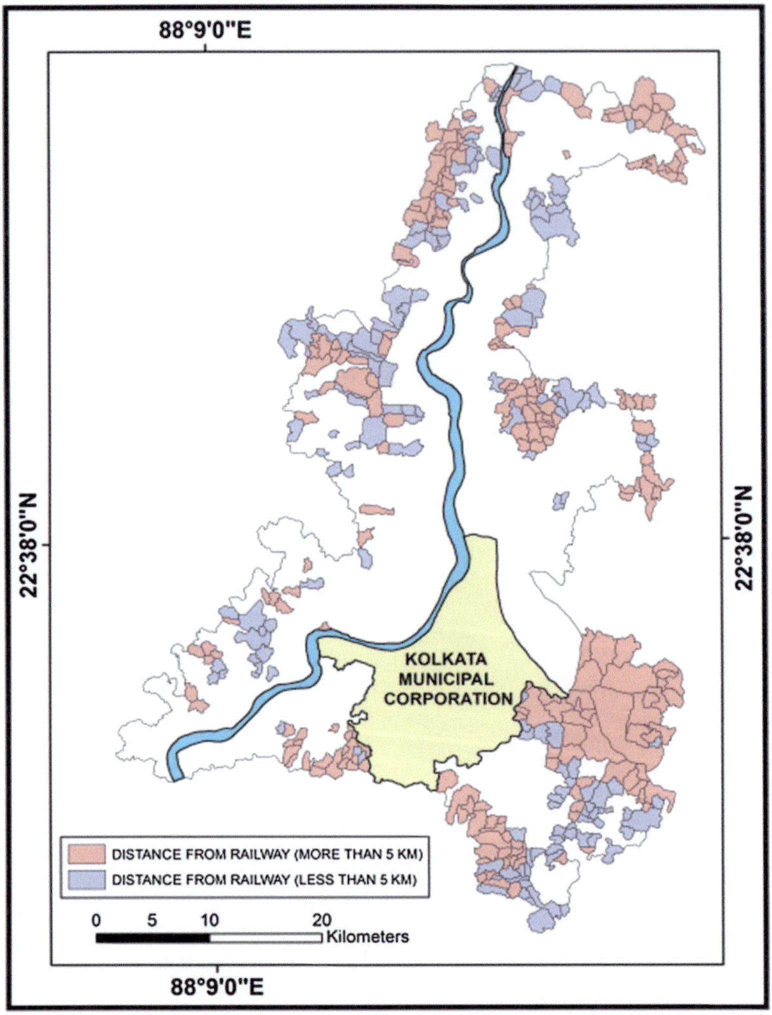

Fig. 12 Distance from the railway line

(rural areas of Chakdaha block of Nadia district), central east (rural areas of the Barasat block of North 24 Parganas district), southeast (rural areas of Sonarpur and Baruipur block of South 24 Parganas district), and northwestern part (rural areas of Hooghly-Chinsurah block of Hooghly district) of KMA the score is high. So it can be easily understand that these areas are the most suitable zones for the peri-urban agriculture around Kolkata city. Other than these above-mentioned areas most of the areas are not suitable for agriculture as these areas are lack of easy accessibility from the other areas, engagement of the people in other activities, etc. Peri-urban agriculture is very important in Kolkata city as most of the vegetables, fishes, and milk used to transport Kolkata city for the daily needs of the city people.

Table 6 Land suitability analysis using multi-criteria decision analysis model

Name of zone	Direction of zone	Percentage of cropland to total cultivated area	Elevation	Percentage of primary workers to total primary workers	Percentage of agricultural area to total cultivated area	Percentage of vegetation to total vegetation	Average water supply and drainage facility	Average market facility	Average bus service and railway facility	Land suitability index (LSI)
Zone I	Northeast	4	5	4	4	5	4	5	5	36
Zone II	Central east	3	4	4	4	4	4	4	5	32
Zone III	Southeast	5	5	5	4	4	4	4	3	34
Zone IV	Southwest	2	2	2	2	3	4	5	4	24
Zone V	Central west	3	2	2	3	4	5	4	5	28
Zone VI	Northwest	5	5	5	4	4	4	4	4	35

5 Conclusion

Kolkata is one of the metropolitan cities in India with 15.89 million populations and scarcity of open land which determines the value of agricultural areas to the people. After independence total number of population in Kolkata has been hugely increased, which is mainly because of influx of refugees from East Pakistan or Bangladesh and it creates extreme pressure on local land resources. This growing population in this city demands more vacant land for the residential purposes which are fixed in supply.

To meet the needs and demand for land, city expands toward the fertile agricultural land and wetlands that causes extensive land transformation in the peripheral areas of the city. Rapid transformation of land possesses a serious threat to agricultural land. With rapid economic growth of the area, the economic structure of the area also shifts from agricultural to non-agricultural based economy. It leads to the additional burden on the present farming conditions, which may further worsen the situation. It means that as the urbanization rises more and more agricultural land is converted to non-agricultural uses, which lead to the reduction of agricultural production. To discourage this conversion the policy-makers should focus on this major issue. In this paper, an attempt has been made to find out the suitable areas for peri-urban agriculture around Kolkata city. From the above analysis, it has been found that agricultural area has been decreased 71% from 1990 to 2015 because of the rapid encroachments of the urban area.

The multi-criteria decision analysis has allocated the land at different scale to find out the suitable places for the peri-urban agriculture of Kolkata city as it is very much needed for the development of Kolkata city. This study identified the level of suitability of the land, particularly for the peri-urban agriculture in spatial dimension. Therefore, the planners and policy-makers of Kolkata city should make different planning strategies for the sustainable development of Kolkata city by preventing those agricultural tracts from the encroachments of the urban areas. Policy-makers and planners should identify the cultivated waste land which can be used for the agricultural industry purposes. The zone with low suitable places for agriculture can be used for dairy farming and fishing activities. As there are no master plans for the planned development of Kolkata city, So Government and policy-makers should make a master plan for the sustainable and planned developments of this city. At last, it can be said that the model is able to allocate the suitable land for the peri-urban agriculture very precisely. This model will help not only the peri-urban planners but also the urban planners, researchers, policy-makers, etc., to take decisions regarding the development of the peri-urban agriculture. If the Government officials and decision-makers of this area follow such a model for preventing peri-urban agriculture of Kolkata city in terms of urban expansion, land allocation, and urban encroachment would be better guided to sustainable and safer.

References

1. Aguilar, A.G., Ward, P.M.: Globalization, regional development, and mega- city expansion in Latin America: analyzing Mexico City's peri-urban hinterland. Cities **20**, 3–21 (2003)
2. Aparicio, N., Villegas, D., Araus, J.L., Casadesus, J., Royo, C.: Relationship between growth traits and spectral vegetation indices in Durum wheat. Crop Sci. **42**(1547), 1555 (2002)
3. Alberti, M., Solera, G., Tsetsi, V.: La cittàsostenibile, F. Angeli (1994)
4. Camagni, R., Capello, R., Nijkamp, P.: Towards sustainable city policy: an economy environment technology nexus. Ecol. Econ. **24**, 1 (1998)
5. Campbell, J.B.: Introduction to Remote Sensing. The Guilford Press (1987)
6. Cavailhès, J., Peeters, D., Sekeris, E., Thisse, J.: The Periurban city: why to live between the suburbs and the country side. Reg. Sci. Urban Econ. **34**, 6 (2004)
7. Estoque, R.C., Murayama, Y.: Intensity and spatial pattern of urban land changes in the megacities of Southeast Asia. Land Use Policy **48**, 213–222 (2015)
8. Guerois M., Pumain D.: Urban Sprawl in France (1950–2000). Franco Angeli (2001)
9. Gao, J., Liu, Y.: Determination of land degradation causes in Tongyu County, Northeast China via land cover change detection. Int. J. Appl. Earth Obs. Geoinf. **12**(1), 9–16 (2010)
10. Liu, F., Zhang, Z., Wang, X.: Forms of urban expansion of Chinese municipalities and provincial capitals, 1970–2013. Rem. Sens. **8**, 930–947 (2016)
11. Lan, Y., Zhang, H., Lacey, R., Hoffmann, W.C., Wu, W.: Development of an integrated sensor and instrumentation system for measuring crop conditions. Agric. Eng. Int. CIGRE J. **11**, 11–15 (2009)
12. Lillesand, T.M., Kiefer, R.W.: Remote Sensing and Image Interpretation, 4th edn. Wiley, New York (1994)
13. Maji, A.K., Nayak, D.C., Krishna, N.D.R., Srinivas, C.V., Kamble, K., Reddy, G.P.O., Velayutham, M.: Soil information system of Arunachal Pradesh in a GIS environment for land use planning. Int. J. Appl. Earth Obs. Geoinf. **3**, 69–77 (2001)
14. Marshall, J.D.: Urban land area and population growth: a new scaling relationship for metropolitan expansion. Urban Stud. **44**, 1889–1904 (2007)
15. Mertes, C., Schneider, A., Sulla-Menashe, D., Tatem, A., Tan, B.: Detecting change in urban areas at continental scales with MODIS data. Rem. Sens. Environ. **158**, 331–347 (2015)
16. Mukhopadhyay, A., Mukherjee, S., Garg, R.D., Ghosh, T.: Spatio-temporal analysis of land use-land cover changes in Delhi using remote sensing and GiS techniques. Int. J. Geomat. Geosci. **4**(1), 213–223 (2013)
17. Shlomo, A., Parent, J., Civco, D.L., Blei, A., Potere, D.: The dimensions of global urban expansion: estimates and projections for all countries, 2000–2050. Progress. Plann. **75**, 53–107 (2011)
18. Thomas, I.L., Benning, V.M., Ching, N.P.: Classification of Remotely Sensed Images. Adam Hilger, Bristol (1987)
19. Weng, Y.C.: Spatiotemporal changes of landscape pattern in response to urbanization. Landscape Urban Plann. **81**(4), 341–353 (2007). https://doi.org/10.1016/j.landurbplan.2007.01.009
20. Xu, H.: A new index for delineating built-up land features in satellite imagery. Int. J. Remote Sens. **29**(14), 4269–4276 (2008). https://doi.org/10.1080/01431160802039957

Security and Privacy Issues in Wireless Sensor Networks

E. Golden Julie and Y. Harold Robinson

Abstract Today, in modern world, technology has improved a lot with wireless communication. Wireless sensor network is normally constructed without any fixed network infrastructure. It is very simple, cheap, self-organized and easy to deploy the nodes in wide range. WSN protocols are mainly designed to focus on energy efficiency and lifetime improvement. Apart from those, WSNs face so many challenges due to limited number of resources available such as memory, computing, space, and bandwidth. The main duty of sensor node is used to collect the information from the environment and transmit the data to the control unit. Due to large number of collection and transmission, they should need to protect from hacker. WSNs need to provide security in all aspects of their architecture reference model. In this chapter, the main contribution includes architecture, attacks and countermeasures.

Keywords Wireless sensor network · Self-organized · Hacker and countermeasures

1 Introduction

The fast growth of technology in wireless communication is widely used in more real-time applications. Normally, wireless communication is broadly divided into two types based on the structure: 1. cellular networks and 2. wireless ad hoc networks.

Basically, cellular network is based on GSM network and IEEE 802.11 wireless LAN. It has fixed infrastructure established communication between mobile nodes and also CDMA-based network for modern mobile computing. It needs good topology for connection establishment to require the network coverage mostly it is in

E. Golden Julie
Department of Computer Science and Engineering, Anna University Regional Campus, Tirunelveli, Tamil Nadu, India
e-mail: goldenjuliephd@gmail.com

Y. Harold Robinson (✉)
Department of Computer Science and Engineering, SCAD College of Engineering and Technology, Tirunelveli, Tamil Nadu, India
e-mail: yhrobinphd@gmail.com

P. K. Pattnaik et al. (eds.), *IoT and Analytics for Agriculture*, Studies in Big Data 63,
https://doi.org/10.1007/978-981-13-9177-4_9

static nature. Wireless ad hoc networks are divided into wireless sensor network and wireless mesh networks. It does not need any fixed infrastructure; it is very easy to deploy and self-organized, and the dynamic nature of nodes frequently changes their state.

In wireless mesh network also like wireless sensor networks but naturally, node deployment is arranged in full mesh or partial mesh that is all the nodes are directly connected to all the nodes in the network. This direct connection supports reliability and flexibility in packet delivery. In the case of wireless sensor network, sensor nodes usually collect the information from the surrounding especially where the wired connection is not possible. This advantage leads to improvement of the sensor network to top of the research, focusing on routing of data, security of path and privacy in data which are some of the challenges. Normal wireless network supports point-to-point or point-to-multipoint data forwarding, but in WSN, its primary goal is to sense, process and forward information to the target.

1.1 Wireless Sensor Network

Wireless sensor network (WSN) is a one of the advance technologies under the combination of electronics and wireless communication which is called micro-electromechanical systems (MEMS) technology.WSN contains hundred or thousand small sensor nodes, each having the ability of collecting data from the surrounding, processing and communicating through the network of nodes.

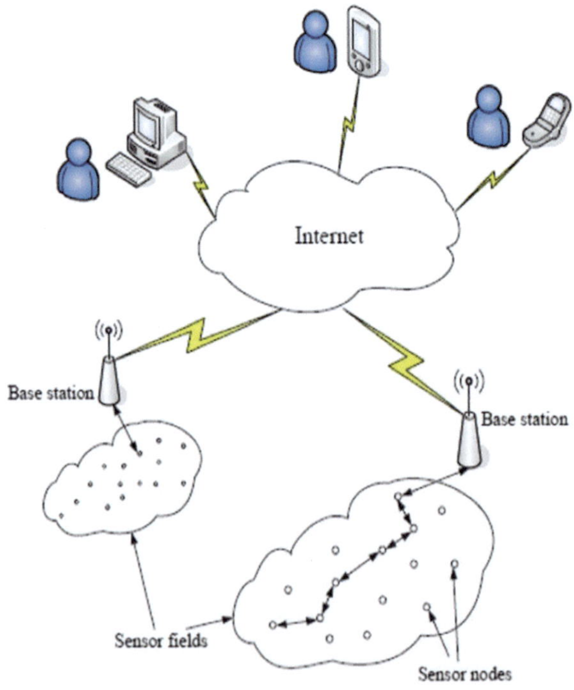

Theoretically, sensor nodes are deployed in the environment to collect data related to pressure, temperature, moisture level and other related parameters. Each sensor node typically has three predefined functionalities: 1. sensing the parameters, 2. processing the raw data (data aggregations, reducing redundancy) and 3. transmitting the data to the sink through wireless link. Some of the sensor nodes themselves act as forwarders to maintain continuous communication in the network. Some special types of sensor nodes have additional functionalities such as forwarding the data received from its neighbouring nodes, data compression and encryption.

Functional view of sensor networks

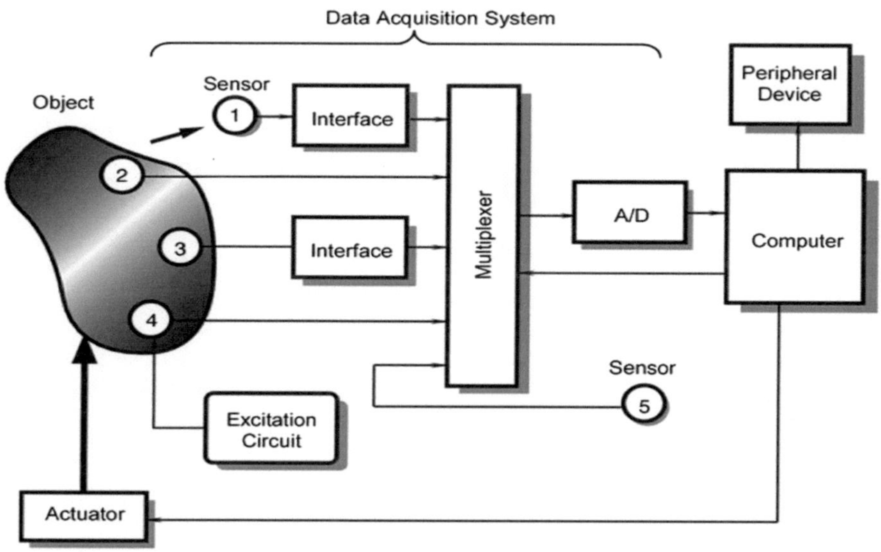

1.1.1 Functional View of Sensor Networks (Jacob Fraden et al. 2010)

In this figure, sensor nodes collect the data as electrical signals that are given into a electronic device (MUX) that is represented as switch or a gate. The operation of the network, the sensor node can connect to the analog-to-digital converter at a time (A/D or ADC). This conversion is based on the output of the sensor node to decide whether it is maybe analog or digital format. Also, it should transmit control signals to the actuator as an object. Samples of the actuators are an electrical motor, a solenoid, a relay, and a gas valve. The functional view of sensor node contains some outputs like alarm, display element, sound recorder, motion recorder, and a number of components that are like filters, sample-and-hold circuits, amplifiers, and so forth.

1.2 WSN Architecture Protocols Stack

Most of the networks are constructed based on Open Systems Interconnection (OSI) model. WSN model contains six layers: application, middleware, transport network data link and physical layer. In addition to six layers, sensor nodes have three planes for entire nodes to manage mobility, power and task. WSN architecture security is one of the major challenges in all six layers.

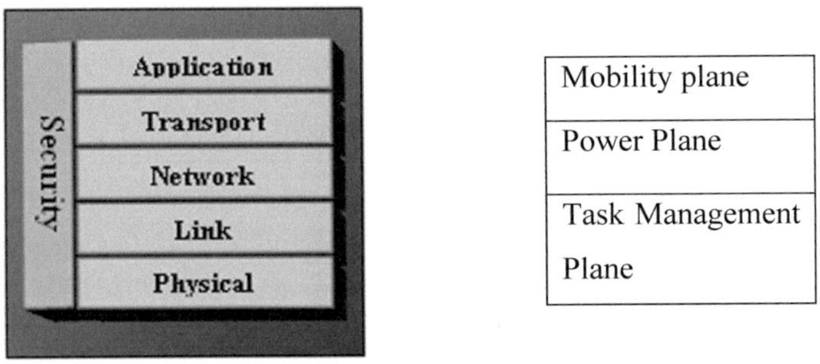

WSN Architecture protocol stack

Application Layer:

It is an abstraction for the main role of the application. It is the final stage of the application. It deals with several processes running simultaneously. It contains message format and order of message sending and receiving by end-users. It is also responsible for providing data required for end-user.

Transport Layer:

The main goal of this layer is data flow and congestion control. Its functionality includes fragmentation and reassembling of the data during transportations. It is also responsible for error control and finding corrupted frame from lower layer to higher layers.

Network layer:

The main performance of network layer is routing and self-configuration. By updating the network information about nodes that are used to find the link failure. The main objective of network layer is to provide optimized network lifetime through frequent updating about position of node.

Data Link layer:

It is an application interface between network and physical layers. Basically, data link layer is divided into two major categories: 1. Medium Access Control (MAC) and 2. Logical Link Control (LLC). In MAC model, efficient channel allocation schemes improve the network lifetime considerably. In LLC, the main function is cyclic redundancy check (CRC) and contains source and destination addresses and sequence number of packet sent and received. Data link layer's main function is multiplexing and data frame detection and efficient sharing of resources between all the nodes in the network.

Physical Layer:

It is a basic layer of the architecture, responsible for translating bits into analog symbol and vice versa. WSN has normal frequency from 915 MHz to 2.4 GHz. It contains frequency selection, modulation symbol synchronization, etc. Normally, binary modulation is applicable for WSN.

Mobility plane:

Normally, sensor node deployment is divided into two categories: 1. static and 2. dynamic. In static model, sensor nodes are placed in fixed position. In dynamic model, sensor nodes were placed in moveable things like animal, vehicle and people. Due to movement of the object, sensor nodes will lead to dynamic topology. Mobility plane is responsible for maintaining list of active neighbouring nodes. On-demand based on the request from the user update neighbouring nodes list and power used through sending and updating data.

Power Plane:

The power efficiency will be achieved through horizontal or vertical layer. The main objective of power management focuses on duty cycle of sensor node based on demand or energy awareness. If the sensor node is placed ideal, it will not be included in the transaction; otherwise, it will be shutdown due to low power.

Task Management Plane:

It is mainly responsible for sharing all the resources available in the network. It maintains the work flow of each layer in the efficient manner with power-aware network. It is responsible for managing tasks like channel allocation, battery power management, etc.

1.3 Application of WSN

The functionalities of the sensor node and networks are mentioned in the previous section, and this section shows the application of wireless sensor network in human life.

It is classified into three major categories of application: 1. environmental monitoring, 2. healthcare monitoring and 3. industrial monitoring.

1.3.1 Environmental Monitoring

Application related to monitoring earth or environment is called environmental sensing.

Air pollution monitoring:

Sensor networks have been placed into the cities to monitor dangerous gases available in the air and find the level of pollution created in the particular date.

Forest fire detection:

Wireless sensor nodes are placed into the dense forest using aircraft to detect the occurrences of fire, causes of fire and spreading direction of fire using temperature and humidity sensors.

Water quality monitoring:

The use of sensor nodes, which are deployed near to dams, rivers and oceans, is to monitor the quality of water by measuring pH level, etc.

Underwater sensor node applications:

Sensor nodes are placed inside the water to detect any changes in the water level and identify the gases available in the oceans.

Natural disaster prevention:

Sensor node deployed in the river side or dam is used to monitor the level of water. The use of this monitoring is to detect flooding in the river. It has features like alarm or message to indicate the occurrences.

1.3.2 Healthcare Monitoring

In modern technology, sensor plays an important role in the field of medical. Three types of sensors are normally used in the field.

1. Implanted Sensor: This type of sensors are placed inside the body to monitor the working nature of the particular organ.
2. Wearable sensor: This type of sensor nodes are placed on the body or close to proximity.
3. Environmentally Embedded: This type of sensor monitors inside the organs working with high-level equipment inside a room connection. Later, this sensor is implemented into IoT for further data processing using Internet.

1.3.3 Industrial Monitoring

In many industries, automatic machines are controlled by using sensor with IoT.

2 Challenges and Issues of WSN

WSN consists of some advantages such as small nodes, low cost, fault tolerance, load balance and network connectivity. These advantages themselves have challenges.

2.1 Energy Constraints

Energy of the WSN depends on three working nature: 1. energy for transducer, 2. energy for sending and receiving data through communication path and 3. energy for microprocessor computation. In WSN environment, transducer energy is higher than receiving energy. So sensor node requires high energy for packet transfer. In the same way, they need energy for each communication like receiving data from multiple nodes. Last they need energy for computing purpose like cryptography security or other security works.

2.2 Memory Limitations

Each sensor node contains small amount of memory for processing collected data. Memory of the sensor node contains flash memory and RAM. Flash memory is used to store coding for sensor node application, and RAM is used to store sensor-collected data, intermediate node information, etc. For example, TinyOS contains about 4 kB. So it is a challenging task for programmer to write code for security algorithms.

2.3 Unreliable Communication

It is one of the major security issues because sensor node communication is based on packet with connectionless protocol. Normally, connectionless protocols are unreliable in nature. Packets may get lost or damaged due to unreliable wireless communication. All the packets are transmitted in broadcast-based transmission. Thus, it leads to higher error rate, higher collusion and increase in retransmission compared to connection-oriented protocols.

2.4 The Lifetime of a Sensor Network

The network lifetime mostly depends on the energy conservation schema. The primary goal of lifespan is based on energy consumption of the communication subsystem. The duty-cycling schema is the best way of saving energy by switching the sensor to ideal when there is no need to monitor the event. The time for which they are switched on is less compared to the idle time. This will reduce the total energy consumption.

2.5 Security

Security of the WSN deals with the following questions: How to protect data? How to safeguard our resources from attack? How to protect our node from misbehaviour nodes?

1. How to protect data?

   Data confidentiality and data integrity provide data protection.

2. How to safeguard our resources from attack?

   Availability, data freshness and self-organization provide safeguard to the packet as well as network.

3. How to protect our node from misbehaviour nodes?

   Secure localization: Time synchronization and authentication are used to identify misbehaviour nodes in the network.

   In many recent applications, security is one of the major issues. The sensed data should be secured up to delivery to sink. For example, military application security is a major issue.

3 Basic Security Schemes in Wireless Sensor Networks

Normally, security focuses on data protection, network protection and location management in WSN. This chapter deals with some security schemes available in the wireless communication environment such as cryptography, steganography and key management.

3.1 Cryptography

In modern computing, cryptography refers to encryption and decryption of data before transferring into a medium. In wireless network environment, cryptography algorithms are not directly applied like wired network. The main divisions of cryptography algorithms are public key cryptography and symmetric key cryptography.

3.1.1 Public Key Cryptography

Public key cryptography mechanism contains two types of keys such as private key and public key for encryption and decryption of message. The proper selection of algorithms based on parameter, optimization and power-aware techniques is very important for WSN. Miller [1] proposes one of the famous public key cryptography algorithm—elliptic curve cryptography (ECC)—which offers distant smaller key size that will reduce processing power and communication overhead. The next public key cryptography algorithm is Diffie—Hellman key agreement protocol proposed by Whitfield Diffie and Martin Hellman in 1976. This scheme is implemented in the network between the two parties without prior knowledge about each other in the network where they establish shared secret key. This key can be used for further communication in the network using symmetric key cipher. This shared secret key communication takes place over insecure communication channel. Rivest et al. [2] proposed (RSA) signatures; it requires large amount of multiplication to perform single operation. It usually needs order of tens of second to perform encryption and decryption operations. It is vulnerable to DoS attacks.

3.1.2 Symmetric Key Cryptography

Symmetric key cryptography mechanism uses only one key for encryption and decryption of message between two nodes. One of the major challenges is sharing a key in a secure manner. Normally, the symmetric key cryptography performance is covered by these two factors embedding data bus size, instruction set will improve the performance of RC4, RC5, IDEA, SHA-1 and MD5 that are some of the symmetric key algorithms used in WSN. Normally, symmetric key cryptography has high speed and low cost than the public key cryptography.

WSN contains low-power nodes in all aspects like energy, memory and computational power. Saleh and Khatib [3] have discussed WSN with encryption mechanisms leads to loss of data increases delay. Moreover, they discussed the type of security mechanisms, which is sufficient for communication, key sharing issues in WSN, issues regarding adding of the new node into the network, time to time key changes is very difficult for dynamic WSN (animal-, vehicle- or human-based WSN).

3.2 Steganography

It is one of the security mechanisms available in the wireless network communication. Kim et al. [4] describe steganography as an art of message hiding using embedding technologies. The main goal of these security mechanisms is embedding the message into image or video carrier without visible and look like ordinary. This type of steganography is very much useful without sender's information and distributed publicly.

Two types of steganography: 1. spatial-domain techniques and 2. transform-domain techniques.

Spatial-Domain Techniques: In this type, information is hidden in the image carrier pixel. It possesses high payload capacity, but it can be attacked by statistical analysis.

Transform-Domain Techniques: In these techniques, the carrier image is exchanged from spatial into transformation like discrete Fourier transform (DFD) and the coefficients of resultant image are changed to implant the top secret information. At last, the image is again exchanged to spatial domain by taking inverse DFD of the resultant image. These methods are having lower payload but are strong against different statistical attacks.

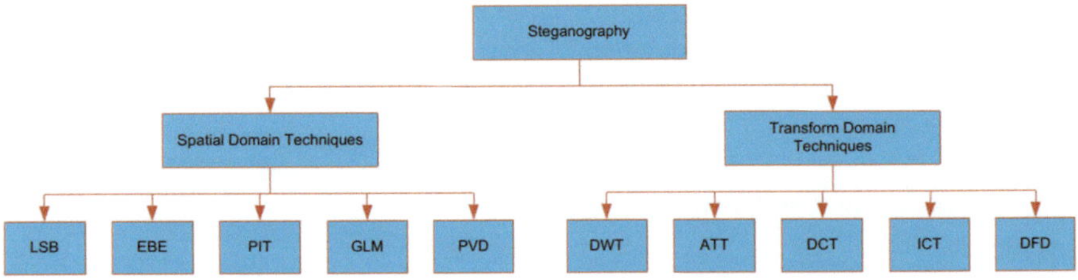

But secure WSN is not directly using steganography because of resources constraints of the sensor node.

4 Security Treats and Defence Techniques for Each Layer of Wireless Sensor Networks

Basically wireless communication is not secure and easy to compromise because of data transmission through waves. In the nature of communication, environment attack is possible. Here, attacks are classified into two types: passive attack and active attack.

Passive attack:

This type attacker is able to interrupt and observe the communication between the sender and receiver, but not intent to modify or tamper. They simply listen to the transaction and do analysis of task, frequency of data, size of data, header position, source and destination location during the transmission, etc.

Active attack:

This type attacker is one of the active participants in all the forms of control and data communication. They can change, remove, rearrange and reply. Some other attackers do node capturing, tamping, etc.

For large-scale WSN, it is very hard to monitor each and every sensor node. Instead of that, we describe layer level attack and protection mechanism applicable all types of networks.

4.1 Physical Layer Attack and Protection Mechanism

Physical layer is attacked by two ways: jamming and tampering.
Jamming:

This type of attacker is more powerful to attack the interface with radio frequencies. Source of the jamming possible to affect entire network or part of the network depends on the power. Jamming sources may be randomly distributed into the environment.

Protection Mechanism for Jamming:

Spread-spectrum communication is best protection mechanism against jamming attack. Frequency-hopping spread spectrum (FHSS) is a scheme of signal transmission by changing carrier among different frequency channels. Without knowing the frequency of transmission, the attacker is incapable to jam the frequency at a particular time. But the attacker knows the range of frequency band, and they may select a large band of frequency to create jam.

Code spreading is one of the techniques against jamming attack. It is commonly used in mobile network because of complex design and energy, not used in WSN. Observation from literature and use of single frequency will susceptible for jamming attacks

Tampering:

It is also one of the physical layer attacks; the attacker can remove important information like data and key of the sensor node. Sometimes, the node may be altered or replaced by some compromised node. This type of compromised node can be controlled by the attacker. This type of attacker can do tamper-proofing of physical package.

Protection Mechanism for Tampering:

Tamper-proofed sensor node is one solution against tampering. However, due to additional cost, mostly nodes will not be tamper-proof. Effective key management schemes must be used for this type of network when sensor nodes are compromised.

4.2 Data Link Layer Attack and Protection Mechanism

The main function of data link layer is multiplexing, data detection, error control and medium access. It ensures reliable connection establishment between different layers. Data link layer is also attacked by two different attacks: resource exhaustion and collisions

Collisions:

Collisions mean two nodes transmit packet simultaneously in the same frequency. This type of attacker creates collisions; they may change the data portion, causing a checksum variance at the target side. The packet will be discarded as worthless.

Protection Mechanism Collisions:

An achievable result of such collisions is based on media access control (MAC) protocols and error correcting codes.

Resource Exhaustion:

This type of attacker uses repeated collisions to cause resource exhaustion. For example, data link layer is responsible for retransmission of corrupted packets. This useless retransmission leads to energy drain.

Protection Mechanism of Resource Exhaustion:

A first technique is applying proper limits to the MAC access control to ignore or reject excess request. This will reduce useless retransmission as well as energy drain.

Second technique—time-division multiplexing—is used to allocate particular time slot for each transmission. This will reduce useless retransmission as well as energy drain.

4.3 Network Layer Attack and Protection Mechanism

The main functions of network layers are routing and power efficiency. Mostly, sensor network is based on data-centric routing (flat), layered approach (hierarchical) and location-based routing

Route information manipulating:

It is one of the direct attacks against routing protocol. Attacker may spoof, alter or replay in order to create traffic in the routing protocols and also create routing loops. The main objective of the type of attackers is to create delay and latency through fake error message, extending and shortening source routes.

Protection Mechanism of Route information manipulating:

The main protection mechanism against this type of attacks is appending message authentication code (MAC). By using MAC code, the receiver can easily identify the spoof or alter packets. In reply attack, appending time stamp or counter are used to identify the reply packet.

Selective forwarding (black hole):

In WSN, always assume multihop communication, all the nodes receive and forward the packet to other node in the network. In this selective forwarding method, attacker will act as malicious nodes and forward some packets and simply drop other packets. One special type of selective forwarder is black hole attacker; this attacker will drop all the received packets.

Protection Mechanism of Selective forwarding (black hole):

The main protection mechanism against this type of attacks is using multiple paths to send data or select alternate path to forward the data.

Sybil attack:

In WSN, collection nodes join together to accomplish the task or subtask. In some situations, one node may use the identity of more nodes in the network. In this type of attack, a node fakes the identity of other legitimate nodes. Sybil attack reduces integrity, security and recourse management.

Protection Mechanism of Sybil attack:

Detection of Sybil attack is not so easy. Radio resource testing is used to detect the presence of Sybil node(s) in the network.

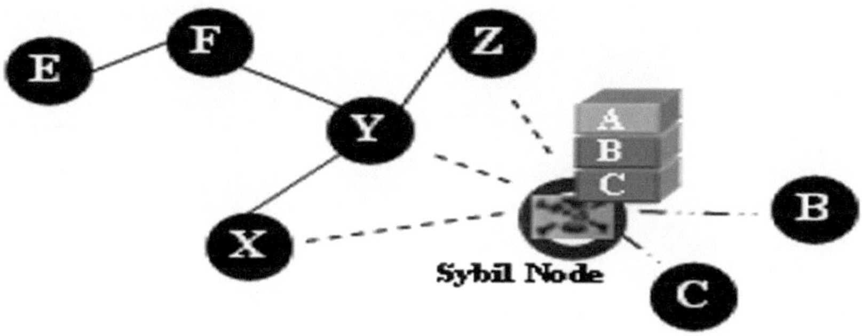

Sinkhole attack:

In sinkhole attack, attacker creates compromised node which acts more reliable node in the network. Particularly, in a flooding-based protocol, the routing updating this compromised node will be chosen as intermediate to route data in the network. Most of the time, this attacker node will be placed near to sink, to get all the data transfer to sink node.

Protection Mechanism of Sinkhole attack:

Protection mechanism of sinkhole attack are authentication and redundancy monitoring of the packet from source to sink and find routing updating periodically.

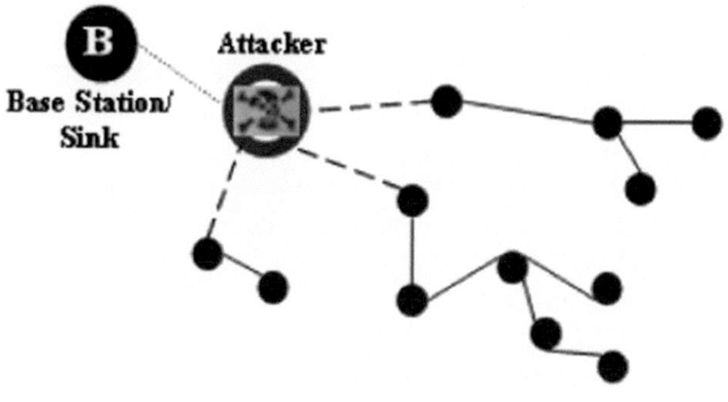

Wormhole attack:

This is one of the critical attacks. In this, attacker will record the data or bit in one location and then tunnel and broadcast in other location. Tunnelling and rebroadcasting is based on selective packet method. Wormhole attack happens in the beginning of route establishment that identifies neighbour node phase itself

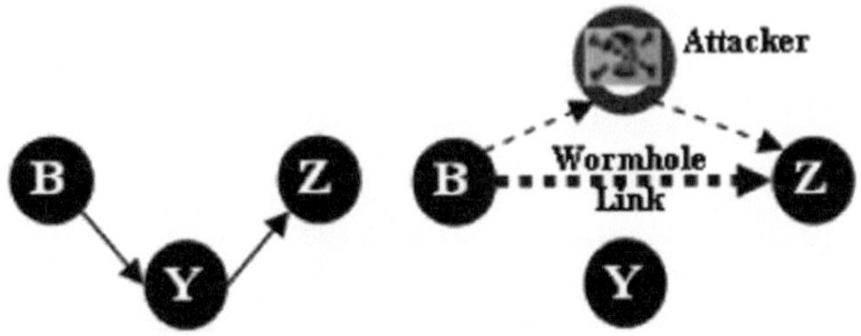

This figure shows the wormhole attack. Attacker is placed in the place of node B and sends request to Z and Y. After getting the reply node Y and Z thinking node B is in our range and update routing table.

Protection Mechanism of Wormhole attack:

Wormhole attack is protected using packet leashes for finding and protecting against it. Geographic leashes and temporal leashes are two types proposed by Hu.

Hello flood attack:

In WSN, connection is established between the nodes in the network through HELLO packet. When sensor node receives the HELLO packet, it assumes that the sender is within the covering range of receiver. Using the high transmitter, attacker can cover large area nodes as a neighbour to get data.

Protection Mechanism of Hello flood attack:

Protection Mechanism of hello flood attack are two-way authentication and three-way handshake protocols.

Transport Layer attack and Protection Mechanism:

Transport layer is responsible for connection established between the nodes.

Flooding:

In wireless sensor network, each node hello flood message to their neighbour identifies the shortest or energy-efficient path in the network. Sometimes, adversary node which is not reachable to sensor node generates flood message in the network and creates unwanted delay, flow of the data transfer and increase of retransmission.

Protection Mechanism of Flooding: Cryptography techniques are used to protect our system from flooding. In this method, nodes share key between them. Ever new encryption key is generated. This method ensures only reachable node can encrypt and verify the message. Limiting connection numbers will protect flooding problems in the network.

Application Layer attack and Protection Mechanism:

It is an abstraction of application, and it produces final product of the application. It deals with message format and end-user product.

Clone attack:

It is most powerful attack in application layer. It compromises fixed number of nodes in the network and creates replication of the nodes into multiple copies. These multiple copies of nodes have legitimate access. This clone node acts as legitimate node to participate into network access.

Protection Mechanism of Clone attack:

It is very hard to detect clone nodes in the network. Pairwise keys are used to prevent clone attack in the network. Effective solution is needed to solve clone attack problem in the wireless sensor network.

5 Secure Routing in WSN

Routing is the process of sharing data between source and destination. This process has two fundamental steps: 1. route discovery (path identification) and 2. forwarding data (route the data). In IP or 3G network routing operations are performed by one special node called routers. Due to resources constrains WSN no special routers are not used in real time. The main objective of routing protocols in WSN should be security and energy-aware protocol design.

5.1 Classification of Routing Protocols in WSN

Flat Network Routing (Data-Centric Routing):

The conventional wireless network like IP network uses node-centric routing. In this node-centric routing, data were sent to unique address schema (IPV4 or IPV6). Based on the query sent by the sink or destination, source sends the data through shortest path to reach destination. In flat network, routing is a data-centric one. In this, each node has the same role and is responsible for sensing the environment. It consists of larger network size and has a huge number of sensor nodes in a network.

Directed diffusion routing protocol is one of the best examples for data-centric routing protocol which is WSN. The basic idea of this algorithm is based on query sensor node response with data. In this routing protocol, sink node broadcasts query like which are the sensors placed in the region area 'y'? Whenever sensor nodes receive the message and send reply to the sink through neighbouring node. Each intermediate node aggregates the data and forwards to sink. Data reaches the sink through some paths. This path will be reinforcement between sink and sensor nodes. Negative reinforcement leads to low data rate.

Hierarchical network routing

Hierarchical routing or cluster-based routing is a layered routing protocol. It utilizes energy in an efficient manner and also improves overall system scalability.

In this, cluster head sends request to the destination through intermediate node. This intermediate node uses aggregation or consolidation of data from destination to sink.

Low-energy adaptive clustering hierarchy (LEACH) is an example for hierarchical routing protocol. In this protocol, sensor node is divided into groups called clusters. In each cluster, one node will be elected as cluster head (CH); the other nodes in the group are called cluster members. The role of cluster head is to collect the data from different clusters and forward that to sink. Rotation of CH is based on time-division multiple access (TDMA).

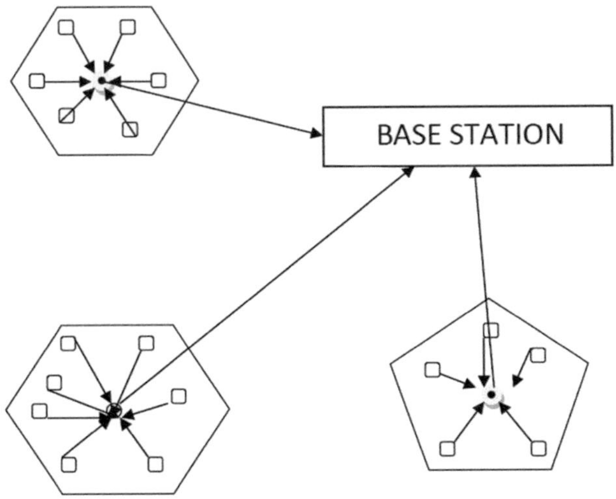

Location-based routing:

In location-based routing, sensor nodes are represented by means of their position, where the nodes are located. It may be available from GPS. Sensor node consists of small GPS receivers to communicate. Physical location of the node is very important for classifying long-range or short-range communication.

Geographic and energy-aware routing (GEAR) is one of the best examples for energy-efficient location-based routing protocol. Sink node selects neighbour based on target region of the sensor network to broadcast packets. It is an improvement algorithm for directed diffusion. The basic idea of GEAR is to reduce number of interests in the whole network instead of sending packet to particular region.

5.2 *Routing Protocols in WSN*

Due to ad hoc nature of sensor node placement, mobility management is needed. This type of WSN is classified into proactive, reactive, hybrid and location-aware routing.

Proactive routing:

Each node in the routing protocol maintains routing information table. This table contains formation about their neighbouring node with hop distance towards destination. Due to node movement, routing table should update frequently to find the path between source and destination. Destination sequenced distance vector is best example for up-to-date routing table maintenances in the wireless network. This update required a lot of energy and frequent update led to overhead are the major drawback of this scheme.

Reactive routing:

This type of routing is based on demand or request–response model. Ad hoc on-demand routing protocol is an example for reactive routing. It contain two phases: first one is route discover, and the second one is response phase. In first phase, route discovery happened based on flooding to find the correct path between source and destination. In response phase, return path is established based on optimal parameters like distance, hop count, etc.

Hybrid routing:

This type of routing contains the combination of proactive and reactive routing protocols. The best example for this type of routing is zone routing protocol (ZRP). In this, node discovery will take place based on proactive method and path establishment based on reactive method.

Location-aware routing:

Geographic adaptive fidelity (GAF) is one of the energy-aware location-based routing protocol.

This routing algorithm increases fidelity by turning off the unnecessary node without affecting level of information gathering. The target areas are covered by GPS-indicated location node to form virtual grid. Based on the routing protocol, the nodes are identified in each process. Thus, it improves energy as well as life of the sensor network.

6 Privacy and Security in Environmental Monitoring System (WSN)

In modern technology, the applications of WSN has been increased rabidly, changes in the environment monitoring. By the use of Internet, the collected data are spread frequently in the worldwide. In this section, we provide system architecture and privacy and security issues with countermeasure available in environmental monitoring.

6.1 System Architecture

Based on the environment, system architecture is classified into centralized, distributed and remote sensing system.

Centralized system:

It contains single processor with number of sensor nodes with single output interface. In this process, data are normally collected from the sensor node and forwarded to control unit. Then, the process forms data analysis based on the application, for example gas detector, smoke or fire detector.

Distributed system:

The distributed system contains three components such as sensor node, intermediate node and central process node in hierarchical form. Each sensor node is used to collect environmental data like temperature, pressure, etc. Intermediate node is used to route the data to the central process unit for further analysis and discussion. Visual sensor network is an example network for this type.

Remote Sensing System:

It is based on signal and image retrieved from artificial satellite- or aircraft-placed sensor node. This type of sensor node is further classified into active sensor and passive sensor. In active sensor node, the signal is passed to the object and the reflected pulse is determined in the object. The passive sensor node only detects quantities produced by the object. It uses predefined wavelength and signal dimension.

Environmental monitoring system can also be divided into three categories based on the geographical extension, such as large scale, regional and localized monitoring.

The large-scale environmental monitoring covers vast area like countries or whole globe, for example, weather monitoring report of the country, ice and snow, desert area measure, etc. The regional monitoring covers particular cities, forest, river, etc. The localized monitoring covers only very local area like lakes, buildings, etc.

6.2 *Environmental Data*

Different types of data can be collected based on the application. Most of environmental data are temperature, movement, speed, chemical condition, audio, videos etc. Normally, collected data are in the form of multidimensional structure. Based on the environmental monitoring system categories, data are classified. In large scale, data collection focuses on the physical parameters which inherit with single phenomenon. Data are collected and aggregated in high frequency to monitor continuously. The region-based monitoring system collects data related to the cities' pollution level, waste collection with GPS to identify the major waste-produced place, etc. Localized monitoring system collects the data of physical parameter using GPS device.

Life cycle of Environmental data:

It contains three macrosteps: 1. collection, 2. storage and 3. publication.

In collection stage, the sensor node is used to collect the data and stored in the sensor/central processing unit. Then, the stored or aggregated data were analysed and the report is generated based on the application. The final stage is publication; in this, the analysed data are published/semipublished as a report. They may be in the form of micro (record reporting)/macro (table reporting).

6.3 Security Risks

In environmental monitoring system, security risk focuses on three categories: 1. damage to the system, 2. violation in the communication channel and 3. unauthorized access to the application.

1. Damage to the system:
   Any attacker attacks the system which affects the physical damage to the system. Let us see one example that focuses on finding good place in city for constructing playground using regional data using GPS. In that time, malicious person hides the sensor placed near to factory A. This person destroys the sensor node placed in the place to avoid monitoring the waste produced by the factory. The same way attacker can attack any part of the environmental life cycle like storage data, published data, etc.
2. Violation in the communication channel;
   Communication channels are used to connect different components of the sensor node in the network. Some passive adversaries are interested only in listening the communication among node and observe the information transmitted in the communication channel. The passive adversary is interested in deleting or modifying data while in the communication. It will affect confidentiality and integrity of the security issues.
3. Unauthorized access:
   Normally, environmental data will be available with user and authorized data owners. Clearly restricted to the person who is not authorized? In the case of data stored data in local server in the control of data owner or third party. In this, server can be considered as trust, or otherwise third party is considered as trust. Based on access control problem, data can be modified or deleted.

6.3.1 Countermeasure for Security Risks

The security issues related to system by handling intrusion detection system and fault tolerance solution can be applied to the situation. The simple solution for availability makes replication in the data to several machines. Communication channel violation can be avoided by using strong encryption techniques suitable for particular environmental monitoring application. Unauthorized access can be solved by any of three techniques: 1. content confidentiality (data privacy is managed), 2. access confidentiality (specific data must be protected) and 3. pattern confidentiality (different accesses aimed at same data must be protected).

6.4 Privacy Risks

Privacy risk is related to allowing the adversary to infer environmental sensitive data from collected data. Inference may be direct or indirect. In direct inference method, data are discovered from stored data. Inference method is based on inference getting relationship between two parameters like illness and pollution of area using geographical environmental monitoring. One example case study of sensitive data privacy risk happens in a hospital. The hospital maintains record of all patients with their diseases. The adversary gets details of sensitive data, and those are published in the media. It is one of the privacy risks associated with monitoring system. User locations are another privacy risk associated with mobile or portable computer. This type of device-based monitoring is called pervasive environmental data. In this sensor, data is collected directly from the environment by the user. If user moves the devices, the location change is shown. In this, adversary gets the detail of user from recorded things like habits, travelled place, etc.

6.4.1 Countermeasure for Privacy Risk

The privacy of environment data can be protected by adapting enhancing privacy solutions like encryption and fragmentation. In this encryption, it is very hard to encrypt all the collected data; instead of that, some of sensitive data are selected and encryption is applied. Next fragment is another technique used to store the data in the form of fragments which leads to increase privacy in collected data. Privacy measure will be applied to all the lifecycle of environmental monitoring. In collection stage, collected data should be encrypted before sending to storage. In storage, data should be protected based on sensitive data column-wise or row-wise.

7 Security Standards in WSN

Security standards contain two types IEEE 802.15.4 and Zigbee. IEEE 802.15.4 deals with physical and MAC layers, and Zigbee deals with network and application layers.
IEEE 802.15.4:

MAC layer provides security and advanced key management support. The MAC layer is used to provide the response to create key, transmit key and stored by higher layers in secure manner.

MAC layer provides security services like access control, encryption and integrity.
Zigbee:

It is an association of companies working together to form standard for reliable, cost-effective, power-efficient wireless sensor network. It is one of the latest techniques used in all commercial and industrial applications. It provides authentication, fairness, encryption and integrity.

Authentication:

Zigbee provides common key for network-level authentication. In the same way, unique key is used between the pair of devices for authentication. By using this method inside and outside, attack may be prevented. But memory cost is higher.

Fairness:

Zigbee contains counters to measure inside and outside data; it increases data fairness. These counters can be reset in every new application.

Encryption:

Zigbee uses 128-bit Advanced Encryption Standard (AES). Network-level and devise-level encryptions are available. Key-based encryption technique is best for network-level as well as device-level encryption.

Integrity:

Zigbee provides three specification options for messages like 32, 64 and 128 bits. By default, 64-bit message integrity is used.

8 Conclusion

This chapter is focused on privacy and security issues in wireless sensor network. The introduction part deals with sensor node and architecture, followed by basic security schemas and issue applications of WSN. Possible attacks and countermeasures were discussed in detail in Sects. 4 and 5. Section 6 deals with environmental monitoring, data security risk and privacy risk which are addressed with possible countermeasures. Finally, security standards available in WSN are discussed [5–53].

References

1. Miller, V.S.: Use of elliptic curves in cryptography. In: Proceedings of the Advances in Cryptology- CRYPTO'85, Santa Barbara, California, USA, Springer LNCS, vol. 218, pp. 417–426 (1986)
2. Rivest, R.L., Shamir, A., Adleman, L.: A method for obtaining digital signatures and public-key cryptosystems. Commun. ACM **26**(1), 96–99 (1983)
3. Saleh, M., Khatib, I.A.: Throughput analysis of WEP Security in Ad Hoc sensor networks. In: Proceeding of the Second International Conference on Innovations in Information Technology (IIT'05), 26–28 September, Dubai (2005)
4. Kim, C.H., Lee, S., Yang, W.I., Lee, H-W., Steganalysis on BPCS steganography. In: Pacific Rim Workshop on Digital Steganography (STEG'03), 3–4 July, Japan (2003)
5. Akyildiz, F., Su, W., Sankarasubramaniam, Y., Cayirci, E.: A survey on sensor networks. IEEE Commun. Mag. **40**(80), 102–114 (2002)
6. Hussain, M., Hussain, M.: A survey of image steganography techniques. Int. J. Adv. Sci. Technol. **54** (2013)
7. Perrig, A., Szewczyk, R., Wen, V., Culler, D., Tygar, J.D.: SPINS: security protocols for sensor networks. Wirel. Netw. **8**(5), 521–534 (2002)

8. Jolly, G., Kuscu, M.C., Kokate, P., Younis, M.: A low-energy key management protocol for wireless sensor networks. In: Proceedings of the Eighth IEEE International Symposium on Computers and Communication (ISCC 2003), vol. 1, pp. 335–340 (2003)
9. Rabaey, J.M., Ammer, J., Karalar, T., Li, S., Otis, B., Sheets, M., Tuan, T.: PicoRadios for wireless sensor networks: the next challenge in ultra-low power design. In: 2002 IEEE International Solid-State Circuits Conference (ISSCC 2002), vol. 1, pp. 200–201, 3–7 Feb 2002
10. Hollar, S.: COTS Dust. Master's Thesis, Electrical Engineering and Computer Science Department, UC Berkeley (2000)
11. Kurak, C., McHugh, J.: A cautionary note on image downgrading in computer security applications. In: Proceedings of the 8th Computer Security Applications Conference, San Antonio, December, pp. 153–159 (1992)
12. Mokowitz, I.S., Longdon, G.E., Chang, L.: A New paradigm hidden in steganography. In: Proceedings of the 2000 Workshop on New Security Paradigms, Ballycotton, County Cork, Ireland, pp. 41–50 (2001)
13. Malan, D.J., Welsh, M., Smith, M.D.: A public-key infrastructure for key distribution in TinyOS based on elliptic curve cryptography. In: Proceedings of the 1st IEEE International Conference on Sensor and Ad Hoc Communications and Networks, Santa Clara, California, Oct 2004
14. Kobiltz, N.: Elliptic curve cryptosystems. Math. Comput. **48**, 203–209 (1987)
15. Ye, F., Lou, H., Lu, S., Zhang, L.: Statistical en-route filtering of injected false data in sensor networks. In: Proceedings of the IEEE INFO COM, Mar 2004
16. Zhu, S., Setia, S., Jajodia, S., Ning, P.: An interleaved hop-by-hop authentication scheme for filtering false data in sensor networks. In: Proceedings of the IEEE Symposium Security and Privacy (2004)
17. Blundo, C., De Santis, A., Herzberg, A., Kutten, S., Vaccaro, U., Yung, M.: Perfectly-secure key distribution for dynamic conferences. In: Proceedings of the Advances in Cryptology (Crypto '92), pp. 471–486, Apr 1992
18. Zhang, W., Subramanian, N., Wang, G.: Lightweight and compromise-resilient message authentication in sensor networks. In: Proceedings of the IEEE INFOCOM, Apr 2008
19. Perrig, R., Tygar, C. J., Song, D.: Efficient authentication and signing of multicast streams over lossy channels. In: Proceedings of the IEEE Symposium Security and Privacy, May 2000
20. Albrecht, M., Gentry, C., Halevi, S., Katz, J.: Attacking Cryptographic Schemes Based on 'Perturbation Polynomials, Report 2009/098 (2009). http://eprint.iacr.org/
21. Chung, T., Roedig, U.: Efficient Key Establishment for Wireless Sensor Networks Using Elliptic Curve Diffie-Hellman (2007)
22. Wacker, A., Knoll, M., Heiber, T., Rothermel, K.: A New Approach for Establishing Pairwise Keys for Securing Wireless Sensor Networks (2005)
23. Zhou, L., Haas, Z.J.: Securing ad hoc networks. IEEE Netw. **13**(6), 24–30 (1999)
24. Strulo, B., Farr, J., Smith, A.: Securing mobile Ad hoc networks—a motivational approach. BT Technol. J. **21**(3), 81–89 (2003)
25. Yang, H., Luo, H., Ye, F., Lu, S., Zhang, L.: Security in mobile Ad Hoc networks: challenges and solutions. IEEE Wirel. Commun. **11**(1), 38–47 (2004)
26. Pathan, A-S.K., Alam, M., Monowar, M., Rabbi, F.: An efficient routing protocol for mobile Ad Hoc networks with neighbor awareness and multicasting. In: Proceedings of the IEEE E-Tech, Karachi, pp. 97–100, 31 July 2004
27. Blackert, W.J., Gregg, D.M., Castner, A.K., Kyle, E.M., Hom, R.L., Jokerst, R.M.: Analyzing interaction between distributed denial of service attacks and mitigation technologies. In: Proceedings of the DARPA Information Survivability Conference and Exposition, vol. 1, pp. 26–36, 22–24 April 2003
28. Wang, B.-T., Schulzrinne, H.: An IP traceback mechanism for reflective DoS attacks. Can. Con. Electric. Comput. Eng. **2**(2–5), 901–904 (2004)
29. Pfleeger, C.P., Pfleeger, S.L.: Security in Computing, 3rd edn., Prentice Hall (2003)
30. Harold Robinson, Y., Rajaram, M.: "Energy-aware multipath routing scheme based on particle swarm optimization in mobile ad hoc networks". Sci. World J. 1–9. (2015). https://doi.org/10.1155/2015/284276

31. Newsome, J., Shi, E., Song, D., Perrig, A.: The Sybil attack in sensor networks: analysis & defenses. In: Proceedings of the Third International Symposium on Information Processing in Sensor Networks. ACM, pp. 259–268 (2004)
32. Julie, E.G., Tamilselvi, S., Robinson, Y. H.: Performance analysis of energy efficient virtual back bone path based cluster routing protocol for WSN. Wireless Pers. Commun. **91**(3), 1171–1189 Springer (2016)
33. Karlof, C., Wagner, D.: Secure routing in wireless sensor networks: attacks and countermeasures. Elsevier's Ad Hoc Network Journal, Special Issue on Sensor Network Applications and Protocols, pp. 293–315, September 2003
34. Hu, Y.-C., Perrig, A., Johnson, D.B.: Packet leashes: a defense against wormhole attacks in wireless networks. In: Twenty-Second Annual Joint Conference of the IEEE Computer and Communications Societies. IEEE INFOCOM 2003, vol. 3, pp. 1976–1986, 30 March–3 April 2003
35. Robinson, Y.H., Julie, E.G., Balaji, S., Ayyasamy, A.: Energy aware clustering scheme in wireless sensor network using neuro-fuzzy approach. Wireless Pers. Commun. **95**(2), 703–721 Springer (2017)
36. Du, W., Deng, J., Han, Y.S., Varshney, P.K.: A pairwise key pre-distribution scheme for wireless sensor networks. In: Proceedings of the 10th ACM Conference on Computer and Communications Security, pp. 42–51 (2003)
37. Oniz, C.C, Tasci, S.E, Savas, E., Ercetin, O., Levi, A.: SeFER: Secure, Flexible and Efficient Routing Protocol for Distributed Sensor Networks, from http://people.sabanciuniv.edu/~levi/SeFER_EWSN.pdf
38. Robinson, Y.H., Julie, E.G., Saravanan, K., Kumar, R.: FD-AOMDV: Fault-tolerant disjoint ad-hoc on-demand multipath distance vector routing algorithm in mobile ad-hoc networks. J. Ambient Intell. Humanized Comput. 1–18. Springer (2017) https://doi.org/10.1007/s12652-018-1126-3
39. Slijepcevic, S., Potkonjak, M., Tsiatsis, V., Zimbeck, S., Srivastava,M.B.: On communication security in wireless ad-hoc sensor networks. In: 11th IEEE International Workshops on Enabling Technologies: Infrastructure for Collaborative Enterprises, pp. 139–144, 10–12 June 2002
40. Balaji, S., Julie, E.G., Robinson, Y.H.: Development of fuzzy based energy efficient cluster routing protocol to increase the lifetime of wireless sensor networks. Mobile Network. Appl. 24(2), 394–406 (2019)
41. Yuan, L., Qu, G.: Design space exploration for energy-efficient secure sensor network. In: Proceedings of the IEEE International Conference on Application-Specific Systems, Architectures and Processors, pp. 88–97, 17–19 July 2002
42. Karlof, C., Sastry, N., Wagner, D.: TinySec: a link layer security architecture for wireless sensor networks. In: Proceedings of the 2nd International Conference on Embedded Networked Sensor Systems, Baltimore, MD, USA, pp. 162–175 (2004)
43. Younis, M., Youssef, M., Arisha, K.: Energy-aware routing in cluster-based sensor networks. In: Proceedings 10th IEEE International Symposium on Modeling, Analysis and Simulation of Computer and Telecommunications Systems, pp. 129–136, 1–16 Oct 2002
44. Y. Harold Robinson & E. Golden Julie, MTPKM: Multipart Trust Based Public Key Management Technique to Reduce Security Vulnerability in Mobile Ad-Hoc Networks, pp. 1–22, Wireless Personal Communications, Springer (2019). https://doi.org/10.1007/s11277-019-06588-4
45. Wood, A.D., Stankovic, J.A., Son, S.H.: JAM: a jammed-area mapping service for sensor networks. In: 24th IEEE Real-Time Systems Symposium, RTSS 2003, pp. 286–297 (2003)
46. Krishnan, R.S., Julie, E.G., Robinson, Y.H., Kumar, R., Abdel-Basset, M., Thong, P.H.: A new algorithm for high power node multicasting in wireless sensor networks. IEEE Acc. **7**, 38584–38592 (2019)
47. Hamid, M.A., Rashid, M-O., Hong, C.S.: Routing security in sensor network: hello flood attack and defense. In: To Appear in IEEE ICNEWS 2006, Dhaka, 2–4 January

48. Robinson, Y.H., Krishnan, R.S., Julie, E.G., Kumar, R., Thong, P.H. Neighbor knowledge-based rebroadcast algorithm for minimizing the routing overhead in Mobile Ad-hoc Networks. Ad Hoc Network. **93**, 101896. Elseiver (2019). https://doi.org/10.1016/j.adhoc.2019.101896
49. Eschenauer, L., Gligor, V.D.: A key-management scheme for distributed sensor networks. In: Proceedings of the ACM CCS'02, pp. 41–47, 18–22 Nov 2002
50. Robinson, Y.H., Julie, E.G., Kumar, R.: Probability-based cluster head selection and fuzzy multipath routing for prolonging lifetime of wireless sensor networks. Peer-to-Peer Network. Appl. 1–15 (2019). https://doi.org/10.1007/s12083-019-00758-8
51. Robinson, Y.H., Julie, E.G.: SMR: a Synchronized Multipath Re-broadcasting mechanism for improving the quality of conversational video service. Wireless Pers. Commun. **104**(3), 1149–1173 (2019)
52. Harold Robinson, Y., Golden Julie, E., Ayyasamy, A, Archana, M.: Cluster based routing in sensor network using soft computing techniques: a survey. Asian J. Res. Soc. Sci. Humanit. **7**(3), 341–360
53. Thanga Aruna, M., Golden Julie, E., Harold Robinson, Y.: A survey on techniques for selection of forwarding node in wireless sensor networks. Int. J. Adv. Comput. Electron. Eng. **2**(4), 24–29 (2017)

A Design of IoT-Based Agricultural System for Optimal Management

Sudipta Sahana, Dharmpal Singh, Souvik Pal and Debabrata Sarddar

Abstract Elegant cultivation practices are the extreme significance for several economies to encourage its expansion and improvement and solve the problem like lack of food and food uncertainty to make sure that people will be happy. Moreover, these efforts typically need huge funds which are beyond the range of the people reach. Therefore, this one is the onus of the researcher of the country to come forward and design a system that will be helpful to the people to achieve the optimal output at low cost. This chapter made an effort to design IoT-based agricultural system using data mining and communication protocol/network for optimal management of agriculture. IoT-based technology makes non-stop message between object/things or between things/object and humans. Data mining and communication protocol will be used as prediction/inference of the knowledge from the hidden data and make the optimal travel of the data with security and recovery, respectively. Therefore, in this chapter, an effort has been made to design an IoT-based agricultural system and discuss the problem and implication of the implementation of these techniques.

Keywords Data mining · Sensor nodes · Topology · Bluetooth low energy

1 Introduction

The Internet of things (IoT), the new technological paradigm, is conquering the entire world by connecting various electronic devices around us. IoT [1] is a con-

S. Sahana (✉) · D. Singh · S. Pal
Department of Computer Science & Engineering, JIS College of Engineering, Kalyani, India
e-mail: ss.jisce@gmail.com

D. Singh
e-mail: dharmpal.singh@jiscollege.ac.in

S. Pal
e-mail: souvikpal22@gmail.com

D. Sarddar
Department of Computer Science & Engineering, University of Kalyani, Kalyani, India
e-mail: dsarddar1@gmail.com

P. K. Pattnaik et al. (eds.), *IoT and Analytics for Agriculture*, Studies in Big Data 63,
https://doi.org/10.1007/978-981-13-9177-4_10

ceptual framework that enables to connect and embed intelligence in devices for establishing a machine-to-machine or machine-to-people or machine-to-object or people-to-machine or people-to-object connection. IoT [2] helps to collect data from the devices with broad range. The twenty-first century has been witnessing the fascination to Internet and gadget as the biggest boon across the world. IoT deals with raw data and processes the data. Hence, in this chapter, we have focused on data mining techniques which are to be implied in agricultural field. In this chapter, we have discussed different data mining techniques, classification, different sensor nodes [3, 4] and functionalities, and how the sensor nodes can be applied in agricultural field.

2 Introduction to Data Mining in Agriculture

In this section, we will discuss the motivation of data mining in agriculture field. We will also deal with different mining techniques related to agriculture.

2.1 Foundation of Agro-Data Mining

Agriculture plays a vital role, and it is the moral fibre of the Indian nation, but under the large areas in India, only one-third of the crop part is irrigated. Due to less production from the agriculture [5] by different reasons, demand of food is increasing day by day. Therefore, this is the onus of the researchers, farmers, agricultural scientists and government to put extra effort and techniques to enhance more production to fulfil the demand. Therefore, as a result, the agricultural data volume increases day by day and involuntary way is required for these data to extract meaningful information and enhance the supply and demand. Due to lack of knowledge, many farmers are not able to save their crops and even do not know how to stop it. Therefore, in this chapter, an effort has been made to use the data mining techniques for predicting the future trends of agricultural processes based on the requirement. Data mining is the process to extract the information from the huge amount of the data and produce the result according to the user need. IoT produces the data from the different sensors and machine which will be used by data mining to produce the result according to the customer need. Different types of data mining techniques are used to predict yield problem using, viz. K-means, K-nearest neighbour (KNN), artificial neural network and support vector machine (SVM). It is not possible to discuss the all the techniques, so here few techniques have been discussed.

2.1.1 Introduction of Data Mining

Data mining is known as knowledge discovery in database, and it can retrieve the meaningful data/inference/knowledge from the huge amount of the data. The data may be from market and any real-life example. But before processing the data, different types of preprocessing techniques have to be applied in the data to make it in proper form for the application of the data mining techniques. During the doling out steps, erudition models/patterns are incidental, by applying the suitable learning system on the pre-processed data. The doling out actions are incorporated in an iterative process, throughout which the majority suitable algorithm and linked parameter values are established (model generation and tuning). Also, the selection of correct algorithm for that is vital task for the data mining. Data mining (DM) method represents an vigorous dare, through their variety and individuality for a firm application. All procedures have actions which can be grouped into three types: preprocessing, processing and post-processing.

2.1.2 Application of Data Mining in Different Domain

The use of data mining is the vast, and it can be used in any area where the extraction of hidden knowledge will be required. It can be used in medical filed to detect the type of disease based on the previous year of data and also able to predict the type of error based on the data. It can be used in market basket analysis to predict the group of buyer and also the more frequent item customer can buy in the future. It can also used in the education to help the admission status of institute and how and why we can attract the student. It can also used in CRM, fraud detection, lie detection and intrusion, customer classification, etc.

2.2 Data Mining Preprocessing Techniques

It has been observed that if the data are not in clean format, then proper extraction of inference is not possible. Therefore, to take the proper data from the data mining, data have to be cleaned in proper format and then apply the techniques and other algorithms. Thus, the demonstration and excellence of data is initially reviewed before operation and analysis. Many data mining preprocessing (data cleaning, data transformation, data integration, data reduction and data discretization) techniques are applied on the data to make it clean. These techniques are used to collect and modified the data taken from the IoT for the further analysis (Fig. 1).

Forms of Data Preprocessing

Data Cleansing

Data purification is the process of detecting and correcting corrupt, incomplete, incorrect, inaccurate, irrelevant imprecise records from a record set, table or database and makes it correct and suitable for use. Therefore, filling in missing values, correcting inconsistent data and identifying outliers and smoothing out noisy data are used in data cleansing part to make the data clean for further use.

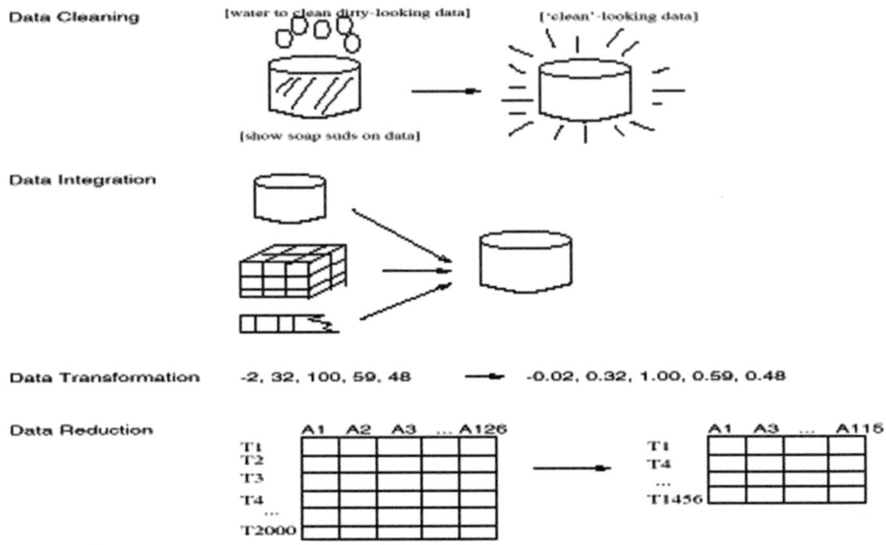

Fig. 1 Data mining process

After cleansing, a data set will turn out to be consistent with other like data sets in the system.

It will also clean the data and make it proper format/whenever the IoT sensors for agriculture received the data.

Data Transformation

Through the appliance of the data transformation techniques (smoothing (removing noise), aggregation, generalization, normalization and attribute/feature construction), data have been consolidated into proper forms.

These processes are used to transform the data and make it proper format for better analysis and make it suitable for learning.

Data Integration

Data integration is used to collect the data and make it in single format for the further use. This is also used to reduce the redundancy data, collected from IoT sensors from the different parts of agriculture. It may be possible that different devices of IoT sensors collect the data of same crops and further increase the data size. So using the data integration, the redundancy data can be minored.

Data Reduction

Databases or data warehouses over and over contain terabytes of data for the analysis and due to this high-dimensionality frequently create problems in the use of algorithms. Therefore, the data reduction techniques (data cube aggregation, dimensionality reduction, removing unimportant attributes, data compression, numerosity reduction, fit data into models and discretization and concept hierarchy generation) are used to reduce the size of the data. Using these processes, the size of huge amount of data will be reduced because IoT device sends the huge data from different sources of agriculture which are required to process in efficient ways using different algorithms.

3 Data Mining Techniques and Their Role in IoT

3.1 Clustering

The Clustering is the way to deals with decision of a structure in a group of unlabelled data. Clustering is an unsupervised learning technique used to make the dispersed data into the similar group. To make the data in similar group, different distance measure functions have to be applied to place the data in appropriate group. This is also useful to find the crops which required same type of the environment soil, temperature, etc., for the growth, and based on the data, we can go for the analysis. For the analysis of the group, different types of distance measure functions have to be used.

Distance Measure

An important component of a clustering algorithm is the distance measure (Euclidean distance, Mahalanobis distance, etc.) between data points. If the components of the data instance vectors are in the same physical units, it is possible that they successfully group similar data instances.

3.1.1 Different Clustering Techniques Used in Agriculture

Clustering Algorithm

K-Means Clustering

k-Means clustering algorithm is unsupervised learning algorithm that is used to solve clustering problem. It assumes k clusters with a priori and defined k centroids for that, one for each cluster. The next step is to form the similar type group of the data based on the data elements which have similar types of properties. At the end, the k-means algorithm does not necessarily discover the most favourable configuration but satisfies the global objective: function minimum. In the agriculture filed, the concept of K-means is used to find the number of crops which required the similar type of environment for the growth. The data taken from the IoT sensors are used for the analysis. K-means is also used for simulating daily precipitations and other weather variables of agriculture.

3.2 Classifications

A classification technique is a systematic approach to classify the data based on the given input. It used decision tree classifiers, rule-based classifiers, neural networks, support vector machines and Naive Bayes classifiers to classify the data in proper group. These techniques used the learning parameter and other constraints to select the data in proper form. Accuracy performance metric is used by classification algorithm to find the accuracy of the result which has defined as:

$$\text{Accuarcy} = \frac{\text{Number of Correct Prediction}}{\text{Total number of Prediction}} = \frac{f_{11} + f_{00}}{f_{11} + f_{10} + f_{01} + f_{00}}$$

Performance metric also defined as follows:

$$\text{Error Rate} = \frac{\text{Number of wrong Prediction}}{\text{Total number of Prediction}} = \frac{f_{10} + f_{01}}{f_{11} + f_{10} + f_{01} + f_{00}}$$

This one is used in agriculture to find the (a) mode of reproduction, (b) mode of pollination, (c) lifespan, (d) growth habit and (e) leaf retention. The method adopted according to the mode of reproduction has classified the crops into two major divisions such as sexual and asexual reproducing plants. The sexual category of plants is developed from a seed or spore after undergoing union of male and female gametes. The typical examples include palms and ferns. The asexual category of plants undergoes reproduction by any vegetative means without the union of the sexual gametes or by apomixes. The crop classifications attempted according to the mode of pollination. These criteria divide plants into three categories such as 1. naturally self-pollinated crops, 2. naturally cross-pollinated crops and 3. both self- and cross-pollinated crops.

3.3 Decision Tree

This technique is used to take the decision based on the data. It initially forms the tree based on the given set of the data, and therefore based on the new data set, it predicts where the data will be belonged. In agriculture part, based on already available data, the decision tree will be formed and this formed decision tree will be used to make the group and decision of agriculture crops. Using the concept of decision tree, agriculture data set estimation can be done based on the other values.

3.4 *Processing Agriculture Data Using Other Techniques*

3.4.1 HDFS

Huge quantity of the data will be composed from the IoT sensors of agriculture data, and these data are required to process in efficient ways. Therefore, Hadoop Distributed File System is used to provide a robust data access as well as high data transfer rate across the highly scalable distributed file system among the Hadoop clusters. In agriculture, it may be possible that IoT sensors are sending the data in fast manner and instant analysis is also required for that; therefore, the concept of HDFC will be used to satisfy the need of that one. It used the concept of MapReducer/R Programming which has discussed in next section.

3.4.2 MapReducer/R Programming

MapReduce concept deals with massive amount of data sets and processed it in parallel or distributed to form the cluster. It is coupled with HDFS to handle huge amount of data. The MapReduce is used (the map stage, the shuffle stage and the reduce stage) to make the huge amount of data into similar groups.

Whenever the data of agriculture are collected from the IoT sensors, then the MapReduce applies its techniques to make it in map stage and therefore shuffle it in such way that proper data will be placed in proper set. Thereafter, it used the reduce stage to reduce the redundancy data to make the whole group to process the data in better ways.

3.4.3 Hadoop

The Hadoop software framework has been designed to chain HDFS and MapReduce to solve many high-end computing problem. But it may face the following furnished challenges, viz. high processing capacity server, enormous time taken, time will be wastage for error and difficulty in program (query building).

Despite these problems, it is used to work on cluster way (maximum of 4500 machines together) and using multiple copies to guarantee dependability of data. It is used to work as parallel with small part of data to save the time which will consume to work in single processing unit. Moreover, if any error occurred at the last stage of long query, then it takes backup data and avoids process loss in case of entity failure.

Furthermore, Hadoop builds the query which is simple which will be code in any language. In the agriculture field, these types of problem may be occurred because the IoT device takes the data which may be required to process in parallel without losing the data. It may be also possible to analyse the data at rapid speed.

4 Type of Communications and Network in IoT

The **Internet of things (IoT)** is the standout technology in today's world. This emerging and technological advancement has captured the attention of people in today's world. The interdependence of computing devices which are ingrained in everyday objects, allowing them to send and receive data, through the usage of Internet, is popularly known as the IoT. The sensors are used in most of the devices that we use in our everyday life, for instance, smartphones, wearable devices, etc. Sensors play a very significant role in our day-to-day life and in IoT. Sensors find its implementation in several applications such as health status (e.g. a heartbeat), air quality, home security. The industrial Internet of things also makes use of sensors to monitor production [6] processes. Hence, it is imperative to know how they work and they can be used to collect information. The fog cloud can be conjugated with IoT to get a new direction in this digital world which helps us in monitoring the agricultural field across the globe.

Generally speaking, the alteration in an environment can be sensed using a sensor. By itself, a sensor is ineffective, but it finds a huge role to play once it is installed in an electronic system. A physical phenomenon (like temperature, pressure, and so on) is determined and transformed into an electric signal with the help of a sensor. A good sensor should have these basic features:

- Highly sensitive to the phenomenon that it senses
- Should not be responsive to other physical phenomena
- Should not modify the measured phenomenon during the measurement process.

4.1 Types of Sensors Used in IoT Network

Sensor is the equipment which can be used for the detection of events or environmental changes. The information can be sent to other electronics, most commonly computer processor, with the help of a sensor. It also involves the activities of process monitoring, including specifying measurements and extracting information. The first thing that comes in peoples' mind is the sensors, when they tend to think about IoT.

These tiny devices are so cheap so that they help in keeping the IoT deployment costs low and an innumerable use cases can be enabled. There is a specific type of sensor required for the installation of each and every IoT. There may be several sensors that may be installed in fields to monitor the agriculture remotely. Let us have a look at each of those sensors (Fig. 2).

Temperature Sensors

Ranging from the factory to agricultural fields, almost all IoT domains make use of the temperature sensors. In manufacturing, the temperature of a machine can be constantly monitored and measured by these types of sensors, to guarantee the fact

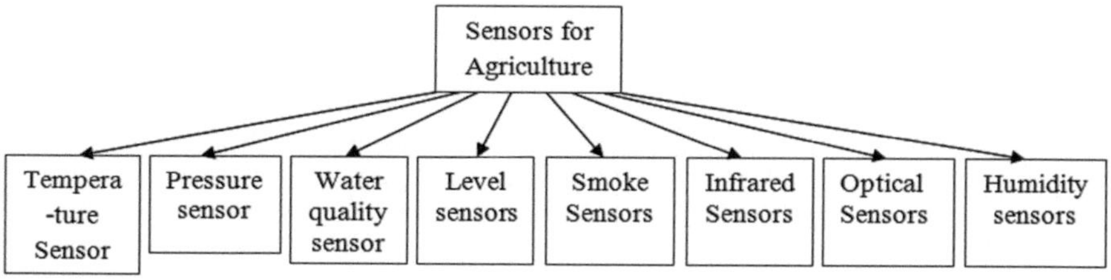

Fig. 2 Classification sensors for agriculture

that it lies within a safe and secure threshold. The output can be maximized on the farm, where the soil temperature, water temperature and plants' temperature can all be tracked by the sensors. The installation of different devices required the usage of these types of sensors. In fact, the invention of IoT has resulted in a large number of applications and implementations of sensors in a wide range of devices. Several applications meant for controlling the environment were based on implementations of these types of sensors. These applications may include A/C control, refrigerators, etc. In fact, they have found a great role to play in manufacturing processes, farming and health industry, etc., with the advent of IoT. There is a requirement of particular surrounding temperature and device temperature, in many machines in the manufacturing process. The manufacturing process always remains ideal with this kind of exercise.

Pressure Sensors

Sensing pressure and converting the same into an electric signal can be attributed to a tool, which is known as the pressure sensor. The extent of pressure that is put in serves as an important parameter for gauging the amount. The course of water by way of pipes and notification of the concerned individual, when there is a need to settle something, can all be done with the help of pressure sensors. Determining force and altitude of a smart vehicles and aircraft also requires the usage of these types of sensors. Moreover, these sensors can also monitor the pressure of the spraying machine used for pest control, and based on the pressure, it will switch on or off the machine.

The liquid or other forms of pressure are some of the aspects on which plenty of devices can rely upon. The IoT systems used for monitoring devices that are pressure driven can also be created with these kinds of sensors. Notifications can be sent to the system manager by the device, regarding any problems that should be sorted out.

Water Quality Sensors

The standard of water and supervision of ion in water distribution systems can all be determined with the help of water quality sensors. We use water in our everyday life for a variety of purposes. A significant role is being played as the standard of water for different purposes can be supervised with the help of these sensors. The water quality sensors are used in some of the common purposes such as satellite farming, improving the water quality and detection of rainwater.

Some frequently used water sensors are listed below:

- ***Chlorine Residual Sensor***: The measurement of low level of chlorine (i.e. residual chlorine, monochloramine and other forms) in water is performed by this sensor, and it is commonly used as germicide, bactericide, etc., thanks to efficiency.
- ***Organic Carbon Sensor***: The organic element in water can be gauged with the implementation of this kind of sensor.
- ***Turbidity Sensor***: The calculation of solids those are suspended in water is performed by these sensors; River and stream gauging, wastewater and sewage measurement make use of these kinds of sensors.
- ***Conductivity Sensor***: The quantity of conductivity is observed organizational processes to get data related to total ionic concentrations (i.e. dissolved compounds) in solutions of water.
- ***pH Sensor***: pH sensors can be implemented in calculating the level of pH in dissolved water, which specifies the amount of acidity or basicity.
- ***Oxygen Reduction Potential Sensor***: The oxygen reduction potential (ORP) gauging gives an understanding of the quantity of reactions related to oxidation/reduction happening in the solution.

Level Sensors

The measurement of quantity of fluids which also include semi-liquid mixture, granular materials and upper surface displaying powders, etc., can be gauged with the help of level sensors. Waste management and recycling process make use of these types of sensors. We may also use this sensor to maintain the suitable water level in agricultural field for different crops like paddy. These kinds of sensors can also be implemented in gauging the volume and size of tank, gas gauging, inventory of liquid assets, etc. Businesses can be rationalized in an improved way, as all the major data are collected by the sensors every now and then. These sensors are also of immense importance to the administrators as the amount of liquid to be disseminated can be identified. These sensors also check for the enhancement and speed-up of the manufacturing processes.

The primary types of measurement are listed below:

- ***Point level sensors***: Point level sensors are mainly used in the identification of certain standards, and the user gets notified once the sensing object exceeds or falls below that standard. A single device makes use of the point level sensors so that an alarm or trigger can be generated.
- ***Continuous level sensor***: The extent of fluid or dry material within a certain scale can be supervised with these types of sensors. The output gets generated where the scale is specified. The fuel level display in the vehicle is one such example.

Smoke Sensors

There are some amazing use cases, which were created, once these sensors were consolidated and combined with the solutions of IoT industrial automation. They are of immense importance in some applications where there is a huge chance of fire and gas leakage, for instance, HVAC, supervision of construction site and the organizational units.

The consolidation and combination of these sensors with an industrial IoT solution have resulted in the prevention of minor gas leakages and some negligible breaking out of fire. Huge calamities can also be averted with these types of sensors. Common types of smoke sensors are present. These sensors check for the presence of fumes, gases and blaze encircling the fields. Optical or the physical process or using of both the methods is mainly used in detecting them.

- ***Optical smoke sensor (Photoelectric)***: A famous technique, called the light scattering principle, is used by these sensors to activate the occupants.
- ***Ionization smoke sensor***: The type of chemistry and the ionization principle are the basic parameters used by these sensors for identifying the molecules which results in a trigger alarm being generated. This sensor is of immense importance in the field of agriculture to monitor the environment.

Infrared Sensors

Infrared radiation can be released and identified, which helps these types of sensors in gauging the characteristics of the environments. The measurement of the heat being emitted by the objects is mainly performed by this sensor. It helps in identifying the pest affecting the crop during cultivation. These sensors can be implemented in the detection of human presence and are deployed in military applications. The implementation of IoT in different healthcare fields is done by these sensors since supervision of flow of blood and blood pressure has become a whole lot easier. A huge range of sophisticated devices like smart watches and smartphones, etc., make use of these kinds of sensors. These sensors find its usage in a whole lot of applications. It can also be implemented as an important tool for ensuring high-end household security.

Optical Sensors

The physical properties of light beams are characterized by these sensors. They enable the conversion of light beams into electrical signals. IoT professionals appreciate these types of sensors, as they can be implemented for identifying different things at the same time. This sensor makes use of a technology where electromagnetic energy can be supervised all the time. This is the reason why these sensors are used in versatile domains. Hence, it has become a whole lot easier nowadays to keep track of the changes in the environment and at the same time ensuring the safety of their employees.

Humidity Sensors

Humidity can be defined as the quantity of water vapour that is present in the atmosphere. One of the significant aspects of humidity is the relative humidity. Humidity sensors are of immense importance in the field of agriculture to check the humidity of the soil. These sensors find a great use in production and construction methods as well as in agricultural domains including cultivation. These sensors enable the total process to be operated in an effective and efficient manner. These sensors can be implemented in organizational domains, business firms and residential areas for a variety of purposes. In other words, these sensors can be put to purpose in versatile domains.

Networking Topologies

The topology of the networks is one of the key considerations while designing IoT networks. Star network and mesh network are the most commonly used networks in constructing the IoT networks. In a star network, there is a point-to-point connection between each network host and the central hub. A Wi-fi network can be a typical example of a star network. This type of network can be featured by the transmission of large blocks of data, speed and fast response time. On the other hand, a limited number of nodes can be supported by a star network that uses Bluetooth. It is highly expensive to scale and build redundancy within a star network. In a mesh network, there is an interconnection between each and every node. Some of the nodes play the role of Internet gateways, and data are transferred into and out of the local proprietary network to the Internet. The communication speed is quite slow; a large number of small hops can be used for the process of communication. The design of mesh networks is more complex than that of star networks. Large number of interior ways for movement is provided by mesh networks. Minor quantities of power for transmission and computation are also required. Nodes can also be added without any hindrance. Hence, mesh topologies are quite cheaper as compared to star topologies.

4.2 Types of Networks

Personal area networks (PAN) are generally wireless and can operate to an extent of some metres. All the useful accessories, including most of the electronic gadgets and electrical appliances, all come under the category of common wireless PAN. These devices generally exhibit small radio communication power and operate on small batteries. Local area networks (LANs) are either wired or wireless or a combination of both wired and wireless. A range up to 100 m can be usually covered in wireless LANs. The home Wi-fi network can be a typical example. Neighbourhood area networks are generally wireless and can operate over more than 20 km. The high-power standards are used by them. They also transfer relatively low data blocks. A smart grid [7] wireless network can come under the category of neighbourhood area network.

4.3 Fog Cloud Computing in Agriculture Platform

The groundbreaking service process employs the recently evolved technologies so that the current service models can be upgraded. The IoT's cloud computing platform can form a basis for constructing the service process. It is utilized for quality improvement of the consolidation and combination of the present cloud-to-physical network. The IoT's computing speed can also be obtained in an effective and better way. Previously, the cloud computing resources and parameters were determined based on cloud computing's high-speed computing or single object's remote operations.

The decision-making can be facilitated by the paradigms of the IoT, which are utilized as resources. A distributed model based on edge and fog computing [8] paradigms is used to implement the IoT architecture, operating rules and smart processes. These technologies are utilized in proposing the communication architecture. The farmers are supported to a great extent in building current and new smart systems. The large-area data collection and analysis can make use of this with the implementation of cloud integration. The farmland is allowed with minimum network information resources to be consolidated and mechanized, which include agricultural monitoring automation, pest management image analysis [6] and monitoring, which are utilized for solving the predicament of large-area automation construction. Different decision trees can be constructed to mechanize the setup, constructed by the farmer, and can be employed easily.

5 Wireless Communication Protocols

However, a number of inventive transforms are occurring with the help of wireless communication protocols. They are categorized depending on the following properties related to operation:

- Data transfer blocks' size
- Connectivity range
- Requirements for power
- Network topology.

Bluetooth Low Energy, Bluetooth Smart

The field of computing has been boosted with the advent of a limited data communication, which is Bluetooth. It can be used for wide range of products linking to IoT via smartphones in a number of cases. For IoT product Bluetooth Low Energy or Bluetooth Smart is very much useful as Bluetooth has much higher power consumption.

In contrast to Bluetooth, file transfers cannot make use of Bluetooth Smart as the size of data packet is quite smaller. As per the forecasts of the industry 90% of Bluetooth-enabled smartphones will be Bluetooth Smart ready soon.

- 2.4 GHz of frequency
- Less than 150 m range
- 1Mbps of data rate.

Zigbee

The IEEE802.15.4 protocol, an industry-standard wireless communication technique, forms the basis for Zigbee and a variety of industrial profiles. This technique is very much useful for applications that require minimum transmission of data at low transmission rate within a range of 100 m, in a building or home. There are a lot of benefits for Zigbee in composite systems which require low-power operation, maximum standards of security, high scalability, high node counts and can bear wireless control and sensor networks [9] in IoT applications.

- 2.4 GHz of frequency
- Less than 100 m range
- 250kbps of data rate.

Z-Wave

It is a less-powered, limited rate of data, transmission technology, intended for house mechanization. Mesh topology can be supported with it. Its feature of scalability allows dominance of up to 232 devices. A much simpler protocol is used by it which provides for quicker evolution.

- 900 MHz of frequency
- 30 m range
- Less than 100 kbps of data rate.

LoRaWAN

It facilitates the wide-area network applications which are provided with minimum power supplies involving IoT's mobile communication, smart city [10] and industrial applications. It is specially used for utilization of low power and holds up networks with a wide range of nodes. The rate of data transmission is as low as less than 50 kbps.

- Various frequencies
- 2–5 km urban range, 15 km suburban range
- Less than 50 kbps of data rate.

6 LoWPAN

The abbreviation for 6LoWPAN is IPv6 low-power wireless personal area network. 6LoWPAN is a communication protocol and can be setup in Wi-fi, 802.15.4 sub-1 GHz ISM bands and Ethernet. The IPv6 stack remains a key characteristic. The IPv6 is very much useful in enabling IoT as a number of users are increasing day by day. IPv4 is the predecessor to IPv6. The unique IP address of IPv6 allows any node in planet to be connected to the Internet. This technology can be implemented in a number of useful applications such as house and building mechanization. A low-powered wireless network in the transportation operation facilitates the linking between complex control systems and the nodes.

Thread

The IPv6 network protocol forms the basis for modelling the thread. It can be used in home environment automation. The existing wireless silicon from chip vendors can be used by it. The IEEE802.15.4-based mesh network is also well supported with it. High levels of security and encryption help in boosting the capability of thread where 250 devices can be handled. Wi-fi can be complemented by thread depending on 6LoWPAN. Although much end-user equipment has hugely benefitted from Wi-fi, it has certain limitations when it is used in home automation shop. Users are allowed to run thread on current IEEE802.15.4-enabled devices by an upgradation of software.

- 2.4 GHz of frequency

Wi-fi

The apparent choice of developers is Wi-fi connectivity, particularly inside home surroundings and local area networks (LANs). It allows faster rate of data transmission, and large size of data can be handled. The consumption of power is quite high for a wide variety of IoT applications.

- 2.4 and 5 GHz bands of frequencies
- Approximately 50 m range
- 150–200 Mbps of data rate is typical, 600 Mbps of data rate is maximum, latest 802.11-ac offers 500 Mbps to 1 Gbps of data rate.

Cellular

Cellular GSM, 3G, 4G assist the implementations in IoT which require function over large distances. Cellular is appropriate for large information volume; however, the utilization of cost and power for operating large volumes of transmission of data is on the higher side for a number of IoT operations. This technique is appropriate for sensor-driven, low data projects, transferred over the Internet.

- 900, 1800, 1900, 2100 MHz of frequencies
- 35 km max range for GSM, 200 km max range for HSPA
- Data rate of less than 170 kps GPRS, less than 384 kbps EDGE, less than 2 Mbps UMTS, less than 10 Mbps HSP, 3–10 Mbps LTE.

NFC

NFC is the abbreviation for near field communication. The flow of information between nodes can be well supported by NFC, a minimum-scale high frequency wireless transmission technology. Two electronic nodes are enabled by it. One of the devices is a portable device like smartphone, which helps in communication establishment by bringing them within 4 cm (1.6 in.) range of one another.

Pros and Cons of NFC:
Pros:

- Availability and wide variety of research activities
- Awfully effortless to use
- Value-added services (VAS)
- Well suited with current RFID framework.

Cons:

- Operates in limited distances
- Minimum rate of transmission
- The technology can be a bit expensive for merchant companies.

Uses of NFC:

- Files and data can be shared between nodes
- Data can be extracted by touching smart posters
- Phone can be used as credit/debit/travel card.

7 Conclusion

The Internet of things will be integrated into the agricultural system and has opened the doors for new and pioneering ideas to bring ease and betterment in the lives of both farmers and labours in the near future. Most of the agricultural system may use it to prepare the process of cultivation and crop monitoring system. Present cultivation system must paradigm shift from traditional to integrate IoT-enabled field. That initiation may create reasonable, practical approaches useful to the future IoT agro-system.

References

1. Asseo, I., Johnson, M., Nilsson, B., Chalapathy, N., Costello, T.J.: The internet of things: riding the wave in higher education. Educ. Rev. **July–August**, 11–31 (2016)
2. Cata, M.: Smart university, a new concept in the Internet of Things. In: The Proceedings of 14th RoEduNet International Conference—Networking in Education and Research (RoEduNet NER), pp. 195–197 (2015)
3. Watfa, M.K., Al-Hassanieh, H., Salmen, S.: The road to immortal sensor nodes. In: 2008 International Conference on Intelligent Sensors, Sensor Networks and Information Processing, Sydney, NSW, pp. 523–528 (2008)
4. Vieira, M.A.M., da Cunha, A.B., da Silva, D.C.: Designing wireless sensor nodes. In: Vassiliadis, S., Wong, S., Hämäläinen, T.D. (eds.) Embedded Computer Systems: Architectures, Modeling, and Simulation. SAMOS 2006. Lecture Notes in Computer Science, vol. 4017. Springer, Berlin, Heidelberg (2006)
5. Marsden, T.K., Munton, R.J.C., Whatmore, S.J., Little, J.K.: Strategies for coping in capitalist agriculture: an examination of the responses of farm families in British agriculture. Geoforum **20**(1), 1–14 (1989)
6. Jayas, D.S., Paliwal, J., Visen, N.S.: Review paper (AE—automation and emerging technologies): multi-layer neural networks for image analysis of agricultural products. J. Agric. Eng. Res. **77**(2), 119–128 (2000)
7. Li, L., Xiaoguang, H., Ke, C., Ketai, H.: The applications of WiFi-based wireless sensor network in internet of things and smart grid. In: The Proceedings of 6th IEEE Conference on Industrial Electronics and Applications, pp. 789–793. Beijing (2011)
8. Sahana, S., Bose, R., Sarddar, D.: Fog-based hierarchical search optimization. In: Mandal, J., Sinha, D. (eds.) Social Transformation—Digital Way. CSI 2018. Communications in Computer and Information Science, vol. 836. Springer, Singapore (2018)
9. Potdar, V., Sharif, A., Chang, E.: Wireless sensor networks: a survey. In: 2009 International Conference on Advanced Information Networking and Applications Workshops, Bradford, pp. 636–641 (2009)
10. Sanju, D.D., Subramani, A., Solanki, V.K.: Smart city: IoT based prototype for parking monitoring & parking management system commanded by mobile app. In: The Proceedings of Second International Conference on Research in Intelligent and Computing in Engineering, pp. 341–343. Poland (2017)

Smart and Sustainable Agriculture Through IoT Interventions: Improvisation, Innovation and Implementation—An Exploratory Study

Arindam Chakrabarty and Tagiya Mudang

Abstract From the dawn of civilization, the unending aspiration toward achieving excellence has been the paramount accelerator which is witnessed through different ages, i.e., Stone Age, Bronze Age, Iron Age, age of automation and information supremacy. The world is transforming into massive digital ecosystem. The comprehensive digital value system is being pioneered by developments in IT and ITES. Internet of things (IoT) is the culmination and assimilation of related instruments for sharing real-time data in a collaborative, harmonized and mutually exclusive manner to facilitate optimum decision-making process. In spite of technological advancement, the society survives on primary sector. So this is the high time to capitalize the threshold the technological knowledge into agriculture system so as to optimize resources, minimize losses and ensure achieving the spirit of sustainability. It is also interesting to see how the most advanced technology can be synergized in primitive farming techniques. The European and Latin American countries have been using IoT in agriculture in varied modes, dimensions and levels. These can be exemplified by glimpse of application like farming based on weather projection, real-life count of agriculture produces, real-life estimation for loss due to perishability or expiry, irrigation issues, controlling of infrastructure support for farming activities from distant location, census of cattle, etc. In fact, the concept of IoT is in still nascent stage in India. There are vast opportunities of IoT application in the country since India is primarily an agrarian society and around 60% population are engaged in this profession which contributes around 17% of share in GDP and feeding the elephantine population of the country. This paper would study various sparks of IoT system, its versatile application worldwide and possible intervention in India particularly in agricultural activities. The paper would explore innovative modeling for IoT integra-

A. Chakrabarty (✉)
Department of Management, Rajiv Gandhi University (Central University),
Rono Hills: Doimukh, Itanagar 791112, Arunachal Pradesh, India
e-mail: arindam.chakrabarty@rgu.ac.in

T. Mudang
Department of Humanities & Social Sciences (HSS), North Eastern Regional
Institute of Science & Technology (NERIST), Nirjuli, AP, India
e-mail: rs_mudang_hss@nerist.ac.in

P. K. Pattnaik et al. (eds.), *IoT and Analytics for Agriculture*, Studies in Big Data 63,
https://doi.org/10.1007/978-981-13-9177-4_11

tion in agriculture system and its ease of implementation globally with emphasizing on Indian subcontinent.

Keywords IoT · Sustainability agriculture · Innovation · Technological advancement

1 Introduction

Concept of IoT

IoT which is popularly known as **Internet of things (IoT)** is the systematic arrangements of smart devices, equipment, objects, animals or people connected with each other assigned with UIDs (known as unique identifiers) and having the features to transfer data over a network.

Today, in every field, the Internet of things (IoT) is getting recognized and becomes significant for agriculturist, professionals, agencies, etc. to understand the phenomenon of its power to improve business and other activities.

IoT basically consists of four basic architectures.

i. **Power-Efficient system**—The inverse proportion of high performance with low battery consumption gives a significant edge during the design of electronic system.
ii. **Cloud computing**—IoT generates a huge mass of data, and these data collected from the device need to be stored and processed in the server. To process such volume of data, there was an inception of cloud computing to anticipate the need. Cloud computing refers to the provision of different services through the Internet. This phenomenon includes tools and applications in the form of data storage, servers, databases, networking and software.
iii. **Big data**—IoT works on devices like radio-frequency identification (RFID) sensors, smart network mostly real time. These devices are spread across every field to capture a massive flux of big data.
iv. **Network connectivity**: Internet connectivity is like oxygen for IoT without which the device cannot work. In order to communicate, the Internet connectivity is essential where each device, object, etc., is represented by an IP address. The smart devices are connected via Internet in a way that it captures and optimizes every bit of data obtained in everyday life.

Dynamics of IoT

The territory of IoT is limitless and is heading beyond computers, laptops and smart gadgets. Nowadays, it is moving toward linked cars, smart homes, connected wearables, smart cities and smart health care. In the nutshell, it is a connected life. As per Gartner report, it is predicted that by year 2020 the number of connected devices across all technology in the world shall reach 20.6 billion.

Year	No. of connected devices
1990	0.3 Million
1999	90.0 Million
2010	5000 Million
2013	9000 Million
2025	1000 Billion

Source https://www8.hp.com

The above table gives the estimation about the number of devices which get connected in the near future. The study conducted by Hewlett-Packard which shows that these devices will fill the space between physical world and the digital to enhance the value, productivity and quality of life, society and industries.
Dynamics of Agriculture:

S. No.	Various domains of agriculture	Description
1	Agronomy	It is the branch of agriculture which deals with the study of crops and soils for the purpose of production of various crops like food crops, fiber crops, sugar, oilseeds, etc., for the purpose of better food production and to control the diseases
2	Horticulture	This branch of agriculture mainly deals with the productions of fruits, vegetables, flowers, ornamental plants, spices, condiments, beverages, etc.
3	Forestry	The forestry branch deals with production of large-scale cultivation of perennial trees for the purpose of wood, timber, rubber, raw materials for industries, etc.
4	Animal husbandry	This branch of agriculture deals with the practice of raring, breeding and raising livestock in order to provide food for humans and to provide power (draught) and manure for crops
5	Fishery science	The science that deals with practice of breeding and rearing fishes. This may include various marine and non-marine fishes, shrimps, prawns, etc., for the purpose of food, feed and manure
6	Agricultural engineering	This branch of agriculture is a branch of engineering also which deals with farm machinery for preparation of field, cultivation, pre- as well as post-harvest process like soil and water conservation engineering and bio-energy

(continued)

(continued)

S. No.	Various domains of agriculture	Description
7	Home science	Home science branch deals with application and utilization of agricultural produces in an optimize manner so as to provide nutritional security, including value addition and food preparation

2 Sustainable Farming

Sustainability is the most prolific buzzword in the present time. The traditional concept of business narrowly focuses on product, productivity and profit. With little or no concern toward environment, natural resources and ecology. The reason behind this ideology was driven by either the availability of abundance of resources or least awareness about nature or natural resources. As the human races are moving through knowledge age, it has been visualized through series of research outcomes, experimentations, etc., that the earth is passing through at alarming stage. The greed of comfort and increased ease of convenience have mesmerized the world population toward innovative products where the virgin resources of nature have been ruthlessly exploited of which hardly miniscule of them can be reversible back to nature, partly of them can be recyclable, and majority of the material have become wastes that cause pollution and diminish ecological balance. On the other hand, as human societies are growing by numbers, and especially in some Asian countries, the growth rate exceeds the G.P rate. As a result of that, the human endeavors have started to understand the resource scarcity for a long-term perspective. All these limiting constraints have necessitated the emerging concept of sustainability. According to Brundtland Commission Report 1987, sustainability can be defined as: "Sustainable development is the development that meets the needs of the present without compromising the ability of future generations to meet their own needs." Agriculture has changed dramatically, especially since the end of World War II. Food and fiber productivity soared due to new technologies, mechanization, increased chemical use, specialization and government policies that favored maximizing production. Today, sustainable farming becomes vital for the health and welfare of our planet. While modern industrial agriculture is highly productive and can produce a massive amount of plants within a harvest season, it also introduces many damaging and long-term problems that can only be solved through sustainable practices. With the decline of the farming environment around the world today, here are five reasons why sustainable farming is a rational solution to our agricultural needs. Sustainable farming system integrates three main goals—environmental health, economic profitability and social and economic equity. A variety of philosophies, policies and practices have contributed to these goals. People in many different capacities, from farmers to consumers, have shared this vision and contributed to it. Technologies and production approaches that meet ecological environmental development requirements are being used for sustainable agriculture. Sustainable agriculture requires the prevention of ecological crises like major accidents, with strong negative impact on the environment.

IoT intervention in Agriculture

The inception of IoT technology has been a boon in the agriculture domain. The use of IoT enables smart farming which helps in reducing the wastage and increasing productivity through efficient production and operations of agriculture raw materials. The use of IT and devices like robotics, automated hardware, sensors and control system in IoT makes possible for precision farming and becomes popular among the agriculture organizations across the globe. To monitor the fields and agriculture activities, the agriculture drone became helpful gadgets for the farmers. This agriculture drone is easy to use, helps in saving time and monitoring crop health and integrates geographic information system (GIS) mapping.

IoT has played an important role in managing the supply chain of agriculture activities. The RFID and cloud computing technology integrate the data and information of production, distribution and safety of quality in the agricultural supply chain effectively which combine the farmers with the IoT and make transparency of the entire agricultural supply chain process. This helps in monitoring and tracing the quality of agricultural materials. Information of every aspect including the production, procurement, storage, transportation and sale on agriculture supply chain management will be processed by IoT. The system sends the exact requirements in terms of quality and quantity of agricultural products such as pesticides, fertilizers and seeds to appropriate places for meeting the needs of farmers in right price at the right time. As much as environment issues are concerned, IoT-based farming benefits in terms of effective usage of water and treatment. As a result, all of these factors will eventually lead to higher revenue generation. The intervention of IoT in agriculture can be advantageously applied in the sphere of agriculture.

3 Review of Literature

Kidd (2012) stated that the application of Internet of things shall provide the base for future sustainable rate of agriculture to meet the estimated growth in the demand for food, with respect to natural constraints, which delivers system-level benefits in terms of biodiversity at different levels (field, farm and landscape). These system will facilitate the land use management and helps in making decisions in the area of production whether to increase or to decrease based on the data gathered. Such a system will give a basis to support farmers in managing their agronomy practices to satisfy local biodiversity requirements and natural resource limitations. Duan (2012) concluded that IoT cannot only improve the quality of agricultural product, the efficiency of agricultural production and the level of agricultural product supply, but also efficiently solve the emergent dispatch of food in special situation. Kanget et al. (2012) states that IoT system helps recognize the information of things such as environmental changes or crop growth changes through the monitoring of environmental information based on the sensors connected through the Internet. Such information is expected to provide the methods to identify the surrounding environments of things through the sensing, delivery and exchanges between differentiated devices. In addition, this information is likely to enable the search and grouping of physically separated growth environments according to various conditions and their management under various conditions corresponding to environmental or other pertinent changes. Moreover, various sensors connected through the Internet are expected to maximize the quality and efficiency of management by conveying the information on the occurrence of abnormal growth to users or controlling irrigation systems through the transmission of control demands. Patil and Kale [5], in their study, proposed an approach with Internet and wireless communications—remote monitoring system (RMS). According to them, this system received real-time information which allows the farmers to cope up with and even benefit from these changes. Due to highly localized nature of agriculture information specifically distinct conditions, it is very challenging task that needs to provide such insights. The complete updated and historical environmental information is expected to help to achieve efficient management and utilization of resources. Zulkifli and Noor [11] observed that with the introduction of RFID technology and wireless moisture sensor network (WMSN) in the agricultural and farming sector, growing crops and plants can be greatly optimized. WMSN reduces the wiring and piping costs and facilitates installation and maintenance in large areas. Apart from decreasing labor costs and water requirement, the application of smart technology in agriculture is very crucial in terms of production. Hence, the WMSN system apparently accomplishes the most technology to enhance the present irrigation systems. The moisture sensors for the soil are continuously improving and becoming reasonable and suitable for massive deployment in the WMSN applications.

4 Objectives of the Study

i. To explore the nations that are practicing any form of smart farming system by reviewing recent and relevant literatures.
ii. To study the scope and opportunities of IoT interventions in Indian farming system.

5 Research Methodology

The paper is exploratory in nature. The purpose of this paper is to understand and study how Internet of things (IoT) has been used for excelling agricultural practices across the globe. The research study has been carried out using reliable secondary information from referred articles, research outcomes, policy papers, etc. The paper also attempted to showcase how the IoT ecosystem can be incorporated in India farming practices. For developing the conceptual framework, inputs from ground level practitioner have been taken for developing such models.

6 Analysis and Interpretation

Analysis-I
IoT intervention in sustainable farming across the world:

Internet of things (IoT) is the assortment of new age instrument that has initiated Fourth Industrial Revolution. The application of IoT has been emerging in the varied fields of agriculture. The developed nations have been experimenting how best the IoT devices and blockchain technology could be assimilated to improve effectiveness and efficiency of the existing modern farming practices. Various research papers have focused on the multifaceted experiences of different countries through their intensive research work. The following table reflects the set of countries who have been using the smart agricultural practices across the world which have been endorsed by relevant research study.

Smart agricultural practices by different countries

S. No.	Name of the country	Source
1	Israel	Lawhon and Schwartz (2006)
2	Brazil	Maia et al. [2] Pivoto et al. [6]
3	Canada	Steele [8] Stanford et al. [7]
4	China	Chunsheng and Ning (2010) Chen et al. (2014) TongKe [9]
5	South Africa	Evans (2018) Masinde [3]
6	Netherlands	Long et al. [1] Van Gossum et al. [10]
7	Australia	Nukala et al. [4]

Analysis-II
Scope of IoT in Indian farming system.

(i) Irrigation system.

The model of smart irrigation provides the solution for the issues of the irrigation system in the conventional farming system and maintains the optimum conditions for their crops. Crops need sufficient water and ideal temperature to grow in the environment. This will help in improving the quality of the plant and productivity of the field too. With the application of IoT devices like smart phones and remote computers, the worker can access the cloud storage to retrieve the data obtained by the sensor devices. The farmer can control the water supply system with the help of the smart devices and can supply the water requirement directly to the root of the plant. This mechanism will help in weed control in the field since the water supply will be limited to the plant only.

(ii) Weather-adjusted agriculture system.

IOT weather reporting system has application in agriculture as well. Weather reporting system based on Internet of things (IoT) application has the ability to foresee and measure about all climatic conditions such as temperature difference, level of humidity and rain. IoT-enabled weather reporting system has been designed with the accessories like LCD display, Wi-fi modem, transformer, bridge, microcontroller, voltage regulator, microprocessor, sensors, etc.

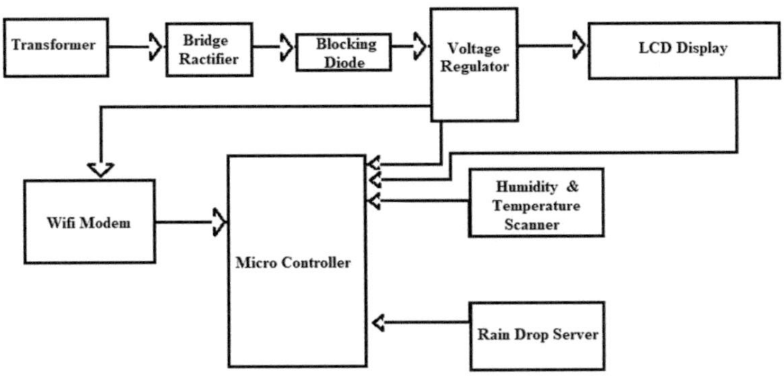

Block Diagram for IoT enabled Weather Reporting System for Farming Excellence

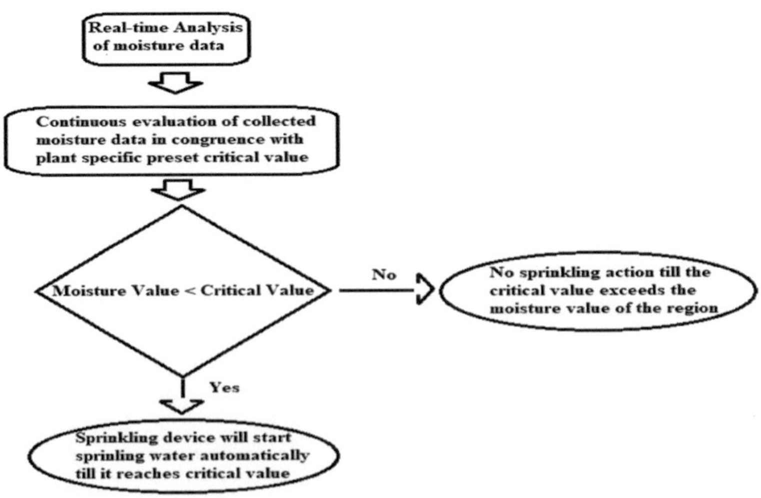

Flowchart of IoT guided smart sprinkling system in Agriculture
Development by Arindam Chakrabarty & Mudang Tagiya

(iii) IoT-controlled sprinkler mechanism.

Smart sprinklers can simplify the garden and agriculture and help in reducing the water consumption. The smart system works on data from the sensors, weather forecasts and crop-care database to ensure the water requirements and provide enough moisture in the right time.
Algorithm for this mechanism.

Step 1: The dedicated moisture sensor shall be designed to detect the moisture in the environment.
Step 2: The smart sensor is installed in the agriculture field accordingly.
Step 3: If the moisture level in the air is less than the critical limit, immediately a message will be sent to the sprinkling system; else, no action will be initiated.
Step 4: The sprinkling system will sprinkle the water immediately as per the instruction.
Step 5: The turning on and off of the sprinkler shall be done based on meeting the moisture value accordingly.

(iv) Estimation and Quality of fruits, pest control, weed control, etc., Using Image Technology.

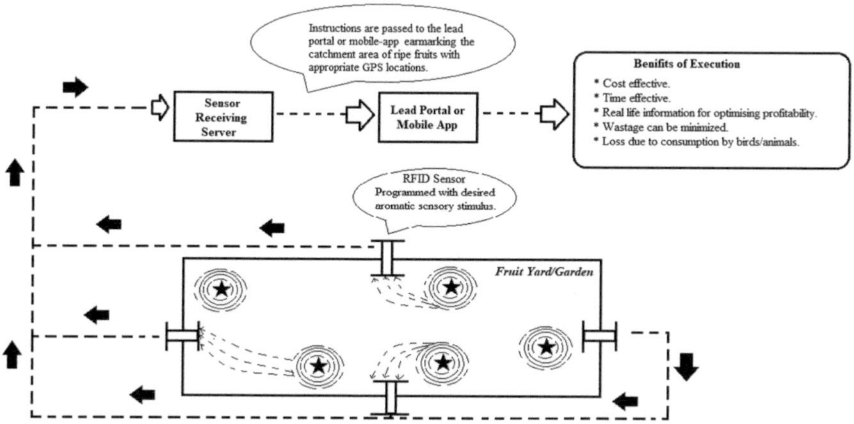

IoT Based Efficient and Sustainable Fruit Plucking Mechanism baseed on RFID sensor using aroma as Stimulus

Algorithm of this mechanism:

Step 1: Dedicated RFID sensor shall be designed which can detect the specific aroma of ripe fruits, and it could transfer to server.
Step 2: RFID sensors are installed in the fruit yard/garden to detect the aroma of the ripe fruits.
Step 3: The high concentration of aromatic stimulus produced by the ripe fruits will be captured by the installed RFID sensors and converted into suitable information about the nature and positioning of the fruits in the large area fruit yard/garden.

Step 4: The captured information is then sent it to the server.
Step 5: The appropriate instructions shall pass to the lead portals or mobile app earmarking the catchment area of ripe fruits with appropriate GPS locations.
Step 6: The farmers may execute the actions and receive the benefits accordingly based on the data retrieved from the system.

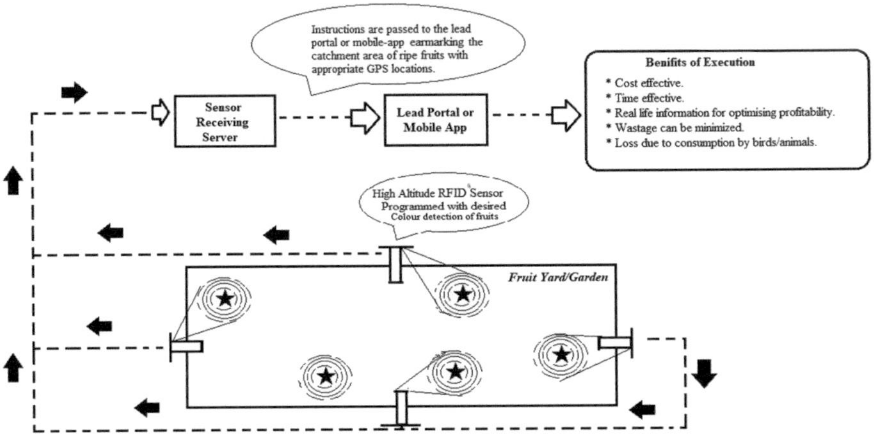

IoT Based Efficient and Sustainable Fruit Plucking Mechanism baseed on RFID sensor using Colour /Pigment as Stimulus

Algorithm of this mechanism:

Step 1: Dedicated RFID sensor shall be designed which can detect the specific color or pigment of ripe fruits, and it could be transferred to server.
Step 2: RFID sensors are installed in the high altitude in the fruit yard/garden to detect the color or pigment of the ripe fruits.
Step 3: The variation or change in the color of the ripe fruits will be captured by the installed RFID sensors and converted into suitable information about the nature and positioning of the fruit in the large area fruit yard/garden.
Step 4: The captured information is then sent it to the server.
Step 5: The appropriate instructions shall be passed to the lead portals or mobile app earmarking the IoT system attached with sensors, and other devices in the field processed the collected data to inform quick action. Remotely monitor for specific pests to understand their activity, location and patterns. This can be done by linking trap stores pond and inform level of specific pest activity followed by monitoring and collecting data to take accurate and quicker anticipation.

Here are four ways, the IoT is revolutionizing pest control in the food industry.

- Prediction and preparation against infestations
- Alerting about the signs of infestation
- It discourages the uses of chemicals
- Handling current issues.

Image analysis to distinguish between the crop and the weed species

i. Fishing—Quality of water is determined by the contents of pH, presence of metals, dissolve oxygen and foreign objects, etc. The most important factor in the production of the fish is the availability of quality water.
The IoT system consists of pond regulator which uses sensory device to check the quality of water in the pond. The sensor captures the activities in the pond and further transfers it to the cloud location for storage. The integrated system is designed in such a way that it will have the capacity for remote operation through sophisticated designed mobile application assessing the data collected by the sensors and also controls the activities of the pond regulator. The regulator will manage the smart feeding system for the fish and the water monitor system for the pond which will help the owner of the pond in managing more than one fish pond from single mobile device. This mechanism will help in reducing the costs of management of fishing breeding activities.

ii. Tracking of domesticated animals–Hilly and tribal areas.
RFID application in animal tracking enables tagging of domestic animals by various countries which helps them in tracking their animals. The tracking is done using radio-frequency identification (RFID) chip which is associated with the tag. This tag is attached to animals (domesticated.) by clamping on their body parts like ear, horn rim, planted inside the animal. Insertion can be done in various places depending on animal (Floyd 2015).

7 Conclusion

The contribution of agriculture in the GDPs of the developed nations has been decreasing in comparison with the service sector and manufacturing. This is simply because of the rapid industrialization, infrastructure development, urbanization and growth of populations. All these factors have been encroaching and shrinking the cultivable lands. On the contrary, the decreasing fertility with alarmingly increasing toxicity in the soil and environment, climate change has adversely impacted on agriculture produce. Within the limiting constraints, the sector needs to be vibrant, value-added and efficient so that the future of human civilization can be sustainable. This paper has attempted to showcase that the assimilation of high-end IT ecosystem with the farming system would be a sustainable solution for the future. The advancement of information technology has been emerging through a series of transformations. Now, it is the age of Internet of things (IoT) and blockchain technology. As per CISCO report, 500,000 million devices are likely to get Internet connected by 2030. The smart devices like sensors, camera, etc., generate data through which IoT examines, combines and delivers insights for decision making. The IoT will be the hub of devices that gathers data in the environment and communicate over the network. As per the recent study, it is calculated that 50% of all US food produce is thrown away. This study essentially has explored various existing and emerging applications of IoT in agriculture sector particularly that suits the nature and aspiration of Indian farming system.

References

1. Long, T.B., Blok, V., Coninx, I.: Barriers to the adoption and diffusion of technological innovations for climate-smart agriculture in Europe: evidence from the Netherlands, France, Switzerland and Italy. J. Clean. Prod. **112**, 9–21 (2016)
2. Maia, R.F., Netto, I., Tran, A.L.H.: Precision agriculture using remote monitoring systems in Brazil. In: 2017 IEEE Global Humanitarian Technology Conference (GHTC), pp. 1–6. IEEE (2017, October)
3. Masinde, M. (2014, July) IoT applications that work for the African continent: Innovation or adoption? In: 2014 12th IEEE International Conference on Industrial Informatics (INDIN), pp. 633–638. IEEE
4. Nukala, R., Panduru, K., Shields, A., Riordan, D., Doody, P., Walsh, J.: Internet of things: a review from 'farm to fork'. In: 2016 27th Irish Signals and Systems Conference (ISSC), pp. 1–6. IEEE (2016, June)
5. Patil, K.A., & Kale, N.R. (2016, December). A model for smart agriculture using IoT. In: 2016 International Conference on Global Trends in Signal Processing, Information Computing and Communication (ICGTSPICC), pp. 543–545. IEEE
6. Pivoto, D., Waquil, P.D., Talamini, E., Finocchio, C.P.S., Dalla Corte, V.F., de Vargas Mores, G.: Scientific development of smart farming technologies and their application in Brazil. Inf. Process. Agric. **5**(1), 21–32 (2018)
7. Stanford, K., Stitt, J., Kellar, J.A., McAllister, T.A.: Traceability in cattle and small ruminants in Canada. Rev. Sci. Tech.-Off. Int. Epizoot. **20**(2), 510–522 (2001)
8. Steele, D.: Analysis of precision agriculture adoption & barriers in western Canada. Final Report (2017)
9. TongKe, F.: Smart agriculture based on cloud computing and IOT. J. Converg. Inf. Technol. **8**(2) (2013)
10. Van Gossum, P., Arts, B., Verheyen, K.: "Smart regulation": Can policy instrument design solve forest policy aims of expansion and sustainability in Flanders and the Netherlands? For. Policy Econ. **16**, 23–34 (2012)
11. Zulkifli, C.Z., Noor, N.N.: Wireless sensor network and internet of things (IoT) solution in agriculture. Pertanika J. Sci. & Technol. **25**(1) (2017)

Websites accessed as on 05.01.2019

12. http://www.gsma.com/newsroom/wp-content/uploads/15625-Connected-Living-Report.pdf
13. https://www.digikey.in/en/articles/techzone/2011/jul/the-five-senses-of-sensors-smell
14. http://h30614.www3.hp.com/collateral/Barcelona2013/presentations/IT3112.pdf

MIX
Papier aus verantwortungsvollen Quellen
Paper from responsible sources
FSC® C105338

If you have any concerns about our products,
you can contact us on
ProductSafety@springernature.com

In case Publisher is established outside the EU,
the EU authorized representative is:
Springer Nature Customer Service Center GmbH
Europaplatz 3, 69115 Heidelberg, Germany

Printed by Libri Plureos GmbH
in Hamburg, Germany